DISASTER RECOVERY

Community-Based Psychosocial
Support in the Aftermath

DISASTER RECOVERY

Community-Based Psychosocial Support in the Aftermath

Edited by

Joseph O. Prewitt Diaz, PhD

AAP | APPLE ACADEMIC PRESS

Apple Academic Press Inc. Apple Academic Press Inc.
3333 Mistwell Crescent 9 Spinnaker Way
Oakville, ON L6L 0A2 Waretown, NJ 08758
Canada USA

© 2018 by Apple Academic Press, Inc.

First issued in paperback 2021

No claim to original U.S. Government works

ISBN 13: 978-1-77-463071-6 (pbk)
ISBN 13: 978-1-77-188631-4 (hbk)

Library and Archives Canada Cataloguing in Publication

Disaster recovery : community-based psychosocial support in the aftermath / edited by Joseph O. Prewitt Diaz, PhD.

Inncludes bibliographical references and index.
Issued in print and electronic formats.
ISBN 978-1-77188-631-4 (hardcover).--ISBN 978-1-315-10268-9 (PDF)

1. Disaster victims--Mental health services--Case studies. 2. Community-based social services--Case studies. 3. Natural disasters--Psychological aspects--Case studies. 4. Natural disasters--Social aspects--Case studies. I. Prewitt-Diaz, Joseph O., editor

RC451.4.D57D57 2018 363.34'8 C2018-901450-4 C2018-901451-2

CIP data on file with US Library of Congress

Apple Academic Press also publishes its books in a variety of electronic formats. Some content that appears in print may not be available in electronic format. For information about Apple Academic Press products, visit our website at **www.appleacademicpress.com** and the CRC Press website at **www.crcpress.com**

CONTENTS

LIST OF CONTRIBUTORS

Dr. Satya Paul Agarwal (in memoriam) was an eminent neurosurgeon, prolific writer, visionary, and secretary general of the Indian Red Cross Society. He served as the Secretary General from 2005 and was also the Chair of the International Federation of Red Cross and Red Crescent Advisory Body on Sustainable Development and Health. He had also served as Director General of Health Services, Government of India, from 1996 to 2005. Dr. Agarwal was a prolific writer who represented India at may international meetings, symposia, and scholarly meetings. He took an interest in psychosocial support in 2006 and had been contributing to the development of the field in India and Southeast Asia. Under Dr. Agarwal's leadership, India had the largest psychosocial support program in the Red Cross movement. In May 2015, Dr. Agarwal corresponded with the Editor about a book on psychosocial support. He reviewed the early chapters of the present work and offered ideas and suggestions for improvement of the theoretical building blocks of the book. In August 2015 he shared with the Editor suggestions for improvement that were later incorporated as part of Chapter 1 of this book. Therefore, he merits posthumous ownership recognition for Chapter 1. His untimely death left a significant void in the completion of this book, his idea.

Cecilie Alessandri, MA, is the psychosocial desk officer for the international department at the French Red Cross. An intercultural psychologist, she also has a master degree in educational sciences. She worked as psychologist, program manager (HIV, mental health and training projects mainly), and trainer in several NGOs (Action Against Hunger, ASMAE, International Medical Corp.) in France and abroad. She manages French Red Cross psychosocial activities in different operational countries and supports dedicated teams to implement and develop psychosocial support for vulnerable persons affected by sicknesses, conflicts, or natural disasters.

Subhasis Bhadra, PhD in Psychiatric Social Work, (NIMHANS, Bangalore) is working as Assistant Professor and Head in the Department of Social Work, Gautam Buddha University. He started his career with his work in intervention in the Gujarat earthquake (2001) and subsequently worked in riots (Gujarat, 2002), tsunamis (2004), Kashmir earthquake and unrest (2005), terrorist attack (Mumbai Serial Train Blast, 2006), cyclone Nargis (Myanmar, 2008), Tsunami in Japan (2011), Himalayan Tsunami (Uttarakhand disaster, 2013) through different organizations, such as Care India, American Red Cross, International Federation of Red Cross, Oxfam India, Action Aid, and International Medical Corps. He supported psychosocial work in a few Asian countries through training and materials development. His research interest includes peace building, life skills education, social work interventions in community, disaster mental health, community and school mental health, and psychosocial support. He has delivered lectures in different institutes/universities on invitation; some of them are School of Social Work, University of Denver, (Skype Lecture); Department of Psychology, University of Indonesia; Department of Education, Tohoku University, Tata Institute of Social Science, Mumbai, Jamia Millia Islamia, New Delhi. Dr. Bhadra is guiding PhD research work on various topics, such as issues of "Dalits," juvenile delinquency, HIV/AIDS, and child care practice in community. Dr. Bhadra is actively engaged in various disaster response programs and lectures on disaster management issues in NIDM (National Institute of Disaster Management), New Delhi, in addition to providing support to different NGOs working on livelihood interventions, community-based care of the orphan children, and psychosocial support in development and disasters. He has published academic articles and book chapters published at the national and international level.

Satyabrata Dash, MD, serves as senior psychiatrist at Apollo Hospitals in Bhubaneshwar, Odisha, India, and is one of the leading mental health care providers in Odisha. Since 2003 Dr. Dash has been involved in the development of psychosocial support programs in India. During the 2004 Southeast Asia Tsunami, he was deployed to the Republic of Maldives where he spent two years as an American Red Cross Psychosocial Delegate. Upon completion of his assignment he was hired as a Psychosocial Delegate by the International Federation of the Red Cross and Red Crescent (IFRC) in Bangladesh. He is credited with developing the training program for the Psychosocial Specialist. His has consulted with the IFRC Reference Center for Psychosocial Support in Copenhagen. He is currently working with Rohingya refugee camps in Bangladesh and the host communities for Action Aid Australia (formerly AustCare). He has worked in post-natural disaster projects on community-based mental health and psychosocial support; designing and implementing country-level programs during emergency as well as during recovery and rehabilitation for five years for the Red Cross Red Crescent Movement covering communities, schools and country level mental health referral systems. Has contributed towards the development of policy documents, strategies and frameworks for countries and organizations. He has experience working in collaboration with partner organizations—governmental, non-governmental and UN agencies including academic institutions and has served as a training facilitator and training course designer in different countries in Asia for different organizations. Dr. Dash has compiled, edited, and adapted training manuals and booklets on mental health and psychosocial aupport and has multiple international research publications and presentations. He is trained in the management of the Psychosocial Support Section of Emergency Response Unit (ERU).

Rashmi Lakshinarayana MD, MPH is Senior Consultant, at EY K.S Institute of Technology, Bengaluru, Karnataka, India. Dr. Lakshminarayana is psychiatrist with extensive experience in disaster mental health. She served as the National Counterpart for the Indian Red Cross Society Disaster Mental Health and Psychosocial Support program that served several million of affected people in Gujarat, Orissa, and Andhra Pradesh and Tamil Nadu. She completed her MSc Public Health in Developing Countries from the LSHTM and holds a postgraduate certificate in Online and Distance Education from the Open University. As a disaster specialized psychiatry, she worked in community-based mental health programs in various states in India with the Indian Red Cross Society, American Red Cross, Oxfam and Action Aid. After completing her degree in public health, she worked with an international perinatal health unit at the Institute of Child Health, London. She helped integrate a maternal mental health component to a cluster randomized controlled trial on reducing neonatal mortality in Jharkhand, India.

Anjana Dayal de Prewitt, MA and Diploma in Humanitarian Diplomacy, serves as the Associate Director for Psychosocial Support and Protection at Save the Children's US Programs. She has worked in several natural disasters and armed conflicts in different parts of the world over the past 14 years. She is currently a member of the Roster Group at the IFRC Reference Center for Psychosocial Support, Copenhagen, Denmark. Most of this time has been with the Red Cross and has included leading community-based psychosocial support programs. As the Country Manager for psychosocial support, she developed the community-based approach for the psychosocial support program in 2005. This program was the flagship for the American Red Cross Tsunami Response. She has provided technical assistance to programs in about twenty countries around the world on themes pertaining to community mobilization, gender, and psychosocial support. She has used qualitative research methodology to conduct conflict analysis, develop national and global guidelines, and design and implement community-based programs, and she published two books. She has also written over ten technical articles for international journals. She speaks English, Hindi, Urdu, and Spanish.

Pramudith D. Rupasinghe, MA, is a psychologist who is specialized in clinical psychology in post-conflict and humanitarian settings. He worked with the Red Cross Movement as well as with the United Nations department of peace operations. He worked in Liberia with United Nations Mission during the West African Ebola crisis and provided his services to the affected populations in the region. Pramudith has been a part of several major disaster responses, including the Haiti earthquake, Great Lake crisis, 2004 South Asian Tsunami, and Nepal earthquake. He has written a book entitled *Footprints in Obscurity: A Living Story*, that provides a profound psychological insight on how childhood dreams often shape one's adulthood. The book presents an image of working in several African countries during civil wars and the Ebola virus.

Jean-Claude Kékoura Zoumanigui is the psychosocial support program supervisor of the French Red Cross, sub-delegation of Macenta, Republic of Guinea. Graduated with a master degree in literature, he also finalized two years' studies of social psychology and worked as a social worker for a project implementing psychosocial support for children and youth severely affected by armed conflicts with the Mano River Women's Peace Network. He also worked for Doctor Without Borders, Belgium, as a psychosocial agent. Nowadays, he supervises psychosocial activities related to Ebola outbreak psychological and social consequences for the French Red Cross in Guinea, Macenta prefecture.

LIST OF ABBREVIATIONS

ADPC	Asian Disaster Preparedness Center
ARC	American Red Cross
AuRC	Australian Red Cross
BCC	behavior change communication
BCDPC	building community disaster preparedness capacity
CBHFA	community-based health and first aid
CBPSS	community-based psychosocial support
CIS	crisis intervention specialist
CIT	crisis intervention technician
CNO	Centre for National Operations
DANIDA	Danish International Development Agency
DHS	demographic and health survey
DMH	disaster mental health
DMHPC	disaster mental health psychosocial care
DRC	Danish Red Cross
DRR	disaster risk reduction
ECHR	European Court of Human Rights
ERU	emergency relief unit
ESB	emotional support brigade
ETC	Ebola treatment center
ETUs	Ebola treatment units
EVD	Ebola virus disease
FAC	family assistance centers
FACT	field assessment and coordination team
FEMA	Federal Emergency Management Agency
GP	general practitioner
HFA	Hyogo Framework for Action
HKRC	Hong Kong Red Cross
IASC	Inter-Agency Standing Committee
ICRC	International Committee of the Red Cross
ICRP	integrated community recovery project

IDP	internally displaced person
IEC	information, education, and communication
IFRC	International Federation of the Red Cross and the Red Crescent
INEE	Inter-Agency Network for Education in Emergencies
IRCS	Indian Red Cross Society
ITN	insecticide-treated net
JRC	Japanese Red Cross
MDGs	Millennium Development Goals
MHPSS	mental health and psychosocial support
NHRC	National Human Rights Commission
NIMH	National Institute of Mental Health
NIMHANS	National Institute of Mental Health & Neuro Sciences
NTL	National Training Laboratories
OFDA	Office of U.S. Foreign Disaster Assistance
PFA	psychological first aid
PHC	primary health care
PHCE	public health in complex emergencies
PNSs	participating national societies
PREMA	Puerto Rico Emergency Management Agency
PSP	psychosocial support program
PSS	planning support system
SLRC	Sri Lanka Red Cross Society
SSCS	social support and counseling services
TOT	Training of the trainers
TRSC	Turkish Red Crescent Society
UDHR	Universal Declaration Human Rights
UN	United Nations
UNDP	United Nations Development Programme
UNICEF	United Nations International Children's Emergency Fund
UNISDR	United Nations Office for Disaster Risk Reduction
UNMEER	United Nations Mission for Ebola Emergency Response
USAID	United States Agency for International Development
WFP	World Food Program
WHO	World Health Organization

ACKNOWLEDGEMENT

We are very pleased to share our practical knowledge about development of community-based psychosocial support. We thank the hundreds of thousands of people in fourteen countries and three continents who provided an immense amount of knowledge about psychosocial support through their participation in the program and pursuing the principle of mutuality, helping us to lead the way in sharing the strategies. We have a special thanks to the American Red Cross and those national societies for providing us the basis to develop a new skill set and a space to implement that knowledge. We thank the IFRC Psychosocial Support Center in Copenhagen for their constant support with materials and timely advice. We acknowledge the many colleagues in so many countries who walked with the authors in the road of development of this platform. Finally, to our families, some too young to understand their input into this process when we were all in the field, now grown up. Many of them want to be psychosocial support assistants and Red Cross workers.

Rejoice, we've conquered!

PREFACE

The last thirty years have brought to the fore the importance of psychosocial support as an integrator and cross cutting theme in disaster response. This book uses case study methodology and practical examples to share how communities can come together, care for themselves, and using its social capital and problem-solving skills can survive and thrive. With a backdrop on the work conducted by the Red Cross, the second part highlights current application and suggest a path to a healthier and more cohesive community.

This book is divided into three distinct parts. The first part of this book composed of three chapters that introduce the theoretical origins of psychosocial support from the optic of the writers. Chapter 1 provides the theoretical building block for a community-based psychosocial support program. Chapter 2 introduces the most recent policy statements and agreements pertaining to mental health and psychosocial support and Chapter 3 introduces the theoretical and practical grounding for community mobilization. Chapter 1 and 2 were prepared by Dr. Prewitt Diaz. Chapter 3 was coauthored by Dr. Prewitt Diaz and Mrs. Dayal de Prewitt.

The second part of the book consists of six chapters. These are the practical chapters that support the model. The precipitant factor was the December 2005 the South-East Asia tsunami impacted 12 countries in the region, the American Red Cross immediately dispatched a group of eight Indian psychosocial experts. Among that group we find four of the authors of this volume (Prewitt Diaz, Dayal de Prewitt, Dash, and Bhatra).

Mrs. Dayal de Prewitt provides a description of the immediate response to the tsunami in the Republic of Maldives. She participated in the translation and contextualization of psychosocial support materials for teachers and children. The third part of her paper presents a case study of the general overview of the development of the long-term psychosocial support response in Sri Lanka

Dr. Bhadra begins his chapter with a description of developing a psychosocial support program after the tsunami in Kanyakumari, Tamil Nadu. The broader part of the chapter explored issues of protection in human rights and it is rooted in the impact of displacement and refugee situations and then ties the IASC-MHPSS guidelines to addressing the needs of the survivors with a visit to diverse populations. He concludes that the community based psychosocial support model is an effective established tool for inclusion of vulnerable population in disaster interventions.

Dr. Dash discusses case studies about the implementation of psychosocial support in the Republic of Maldives, and transferring the technology to Bangladesh. The discussion includes development of materials, capacity building, development of materials, and providing psychological first aid and staff care to the personnel.

The third part of the book is composed of five chapters. These chapters present some of the current issues where psychosocial support is used as a foundation to address new challenges. They are tied together using the IASC-MHPSS guidance as adapted to epidemics and events of mass destruction.

In Chapter 9, Mr. Rupashinghe addresses the psychosocial support response from a macro-level, while serving in a United Nations response unit that coordinated three of the affected countries: Liberia, Sierra-Leone, and Guinea. He concludes that community-based psychosocial support programs founded on local believes and sensitive to the social, cultural, and religious aspects of the affected countries are a vital need.

In Chapter 10, Mr. Zoumanigui and Ms. Alessandri describe the French Red Cross developed a psychosocial support program in a village in Guinea. While this case study is localized, it highlights the importance of the IASC-MHPSS guidance in developing the psychosocial support activities.

In Chapter 11, Dr. Prewitt Diaz presents case studies of events of mass casualty in two schools in the United States and a major fire in India, and reviews the literature and the measures that have been improved to alleviate fear in children and make schools a safer place.

In Chapter 12, Dr. Prewitt Diaz explores the impact of the Zika virus in three Latin American countries. This is the first time that psychosocial support is integrated as part of the medical and preventive community activities. He concludes that the psychosocial interventions have alleviated fear and suggest that an integration of psychosis support in the immediate response

In Chapter 13, Dr. Prewitt Diaz brings to the fore three evidenced informed practices in program development, materials and instruments. He includes best practices that should be followed by practitioners and scholars.

Chapter 14 and 15 address the recent psychological effect of Hurricane Maria 2017 in Puerto Rico. These Chapters written by Dr. Prewitt Diaz who was part of the first contingent of Red cross personnel to arrive to the Island after the hurricane. He highlights the feelings and emotions expressed by the survivors.

ABOUT THE EDITOR

Joseph O. Prewitt Diaz, PhD, serves as the editor for this book. He is a practitioner scholar who has been doing field work and writing about psychosocial support in communities since 1998 in the Caribbean, the Americas, and South Asia. He has authored or co-authored eight books in Spanish and English on the subject and over 40 refereed journal articles. His work was recognized by the American Psychological Association (APA) by his being awarded the International Humanitarian Award in 2008 for assisting in the design and preparation of international guidelines and standards, rapid response, training of staff, program management and implementation, and monitoring of psychosocial support on behalf of the American Red Cross (ARC) in South Asia. He served as the ARC Senior Technical Advisor for Psychosocial Support from 2005 to 2007. This $32 million-dollar program served over 733,700 people and trained over 29,000 PSP technicians over a five-year period. He is currently serving as a Psychosocial Technical Expert with the IFRC Psychosocial Support Reference Center in Copenhagen, Denmark. Disaster Mental Health Advisor for the National Capitol Region of the American Red Cross.

INTRODUCTION

The purpose of this book is to document the evolution of psychosocial support for the last 20 years and in different parts of world, to reach an awareness is the primary task in reestablishing place after a major disaster, a humanitarian crisis or a slowly evolving epidemic. The linkage among all the authors is our relation to the Red Cross movement. So, the works presented herein can be construed as the experience of some field related wisdom achieved by the authors while developing and implementing psychosocial support programs in communities in several parts of the world.

The journey began the first week of December 1998. Hurricane Mitch had hit Central America quite severely. A month before, 1600 ml of rain fell on the Las Casitas volcano causing the highest side to collapse. The huge mudslide killed 1680 people, with over 500 disappeared later to be confirmed dead. The approximate 200,000 m3 of volcanic rock buried two cities: El Porvenir and Rolando Rodriguez.

I had been involved with the American Red Cross, as a disaster mental health volunteer; since I was a native Spanish speaker, I received an invitation from international services to assist in the 12 survivor camps to assist the mental health services currently being provided to the survivors. My background as a psychologist specializing on culturally and linguistically diverse populations in the United States would be of great help, I was told.

The mental health services 6 weeks after the disaster had been relegated to the back burner, with ad hoc services provided by some 20 civil society organization and some INGOs. Most of the services provided were individual counseling and referral to the local hospital for medication. For children, there were recreational activities, several times a day depending on the external help.

We very quickly shifted from providing disaster mental health services to attempting to organize a response that would provide psychological and social services. Our first move was to set up a space in a town near-by to coordinate all services. After conducting the agency mapping, we realized that we had very good resources and that we needed to listen to the people and help them prioritize their needs. Identify the most vulnerable

groups in the community, and organize all to care for themselves (neighbors helping neighbor's concept). We notified American Red Cross International Services that we would not be doing Disaster Mental Health but
rater use an approach that all agreed would be "psychosocial support."
Thus, our first guidance came to light "Salud psicosocial en un desastre
complejo: El efecto del huracan Mitch en Nicaragua" (August 15, 2000)
develop by members of the community and the participating civil society
organizations.

The approach dropped "debriefing" as a preferred tool and began to use
the basic principles of psychological first aid. Safe spaces were developed
for four groups of children by age in each camp, persons who had lost
limb because of the disaster and couldn't work the fields became teachers
and care takers in the safe spaces, women and men had crafts shop with
donated raw materials, and diverse faith groups and their traditional activities were accommodated in a communal tent in each of the refugee camps.
The activity had its challenges that we were able to remedy (1) children had
never used crayons, and so we included activities that would develop fine
and gross motor activities; (2) about half of the children spoke "Wis-Wis"
a Miskito dialect, while instruction was provided in Spanish, we brought in
community members that could help the children in their native language,
and (3) stories that were provided from outside were not congruent with the
cultural context, so we developed story telling competitions, the yield was
a primer for each of the age groups in their native language and contextual to their current situation. The yield was that by drawing, talking, and
writing many people could share their feelings and feel better about it. So,
began the first psychosocial support effort supported by the American Red
Cross in Central America. By mid-1999, I was appointed as the Regional
Program Coordinator for the American Red Cross Regional Delegation
with specific responsibilities for Health, Water and Sanitation, Disaster
Preparedness and Response, and the new field of "psychosocial support."

With this added responsibilities, we began to introduce psychosocial
support into the parlance of disaster response. We began working closely
with the Guatemala Red Cross and introduced a preliminary course on
psychosocial first aid (Primeros Auxilios Psicologicos) and Crisis Intervention Technicians (Tecnicos en Intervencion en Crisis). Both groups had
active participation in the Cubana de Aviacion airplane crash of December
21, 1999. This was the first public response of both groups; within a month
they were a permanent fixture for the Guatemala Emergency Agency
(CONRED). After action reports and suggestions by program participant

and the affected communities let to fine tune the services. The Guatemalan Red Cross adopted psychosocial support (apoyo psicosocial comunitario). By 2002 a methodological guide was developed to standardize training of personnel "Diplomado para la Intervencion en Crisis: Guia Metodologica." The yield of this effort was to identify three primary tasks for psychosocial support: (1) community mapping, (2) psychological first aid, and (3) use of community capital to implement the response with technical and financial assistance from external stakeholders.

On the meantime Dr. Jim Ricca, the Health delegate in Honduras entered a collaborative with the Medical School of the Universidad Nacional de Honduras in August 2000 to build the capacity of 20 doctors in techniques of psychosocial support. The course of study lasted 24 weeks. It provided for three sessions of 15 h each of didactic teaching and 1 week of field work per month. The feedback from the doctors added some additional pieces to psychosocial support. All disaster volunteers in the different Red Cross chapters in the country were trained in psychological first aid. Community leaders and volunteers were also trained in psychological first aid and community mapping.

This knowledge was transferred to El Salvador, where three technicians from the Guatemalan Red Cross, supported by the American Red Cross were sent to the Lempa River basin. This zone is important because its basin is in three countries: (1) El Salvador, (2) Honduras and (3) Guatemala. The three technicians had worked with 16 communities developing the skills of 162 community members in what was called: "the psychosocial support in communities" (damage assessment, conflict resolution, psychological first aid, and referral of people to the health post). The total yield of this exercise is that the volunteers developed a contextualized non-verbal tool that included 187 slides. Two events are worth mentioning about this experience. About noontime on January 13, 2001, a 7.6 earthquake took place, our office was severely damaged, and the fatalities were over 1000 in a landslide in Santa Tecla.

While working with the emergency crews, I continued to call on the radio the volunteers in the Rio Lempa basin. At about my worries were dispelled by the radio message: "this is Mencho, all 62 communities have reported back, we are all saved." A week later three events merit mention: (1) a group of elderly ladies were having problems sleeping and carrying for their daily chores, they were afraid that another earthquake would come and hurt them. (2) In a community meeting we were discussing what would help them feel better. One lady suggested that if there was a way to watch the TV afternoon soaps they would surely calm down. The new procedure was to attach every afternoon for thereon to car batteries so that the older population would feel calm and begin to rebuild their place. (3) All volunteers of the Salvadorian Red Cross were provided with individual Psychological First Aid, and each Chapter had the availed themselves of a "story telling day" where families and volunteers shared with their peer group their feelings about that month.

In September 2002, the program and practices as well as didactic materials were taken from Central America to India. Drs. Gordy Dodge and Jerry A. Jacobs had both worked in the recovery process for the Gujarat Earthquake and laid the ground work for the development of a long-term psychosocial support program. The Indian Red Cross Society had agreed with the report of these colleagues and with the initial ground work that had been set up. Two giants of psychosocial support in India; Drs. R. Srinivasa Murthy and K. Sekar lead a group of Indian Red Cross volunteers and paid staff in a meeting to plan the road that the Indian Red Cross would follow. Both gentlemen were veterans from the many disasters that had occurred in India prior to our arrival. They had responded to the Marathwada earthquake, the Orissa super cyclone, the Gujarat-Kutch earthquake, the Gujarat riots, and the tsunami. Their contribution in terms of manuals, evaluation reports, training of staff and volunteers, and consulting with the American and Indian Red Cross have been very important contributions to the development of psychosocial support response in India and the whole of South Asia.

Furthermore, the collective work of Dr. Murthy and Dr. Sekar in conjunction with Dr. Bhatra and Mrs. Dayal Prewitt, among others leads the National Disaster Management Authority of India to the formal definition of psychosocial support in India.

> Psycho-social support in the context of disasters refers to comprehensive interventions aimed at addressing a wide range of psycho-social problems arising in the aftermath of a disaster. Psycho-social support and mental health services should be considered as a continuum of the interventions in disaster situations. Psycho-social support will comprise of general interventions related to the larger issues of relief work needs, social relationships and harmony to promote or protect psycho-social well-being of the survivors. Mental health services will comprise of interventions aimed at prevention or treatment of psychological symptoms or disorders. These interventions help individuals, families and groups to restore social cohesion and infrastructure along with maintaining their independence and dignity.

The psychosocial program developed by the Indian Red Cross Society trained thousands of volunteer and paid staff in psychosocial support. Among the contribution was the development and contextualized materials, development of training models for community workers, volunteers and Red Cross paid staff, and long term follow-up of communities affected by disasters in Odisha, Gujarat and Tamil Nadu. By 2004 the consortia of government officers, local non-government agencies, and international agencies had reached an agreement and a potential agreement on the definition of psychosocial support presented above.

Indian organizations, specifically OXFAM-India had lead an effort to insert language in the Sphere project that supported psychosocial support. We served as member of this committee in India and subsequently part of the writing group in Geneva. Thus, psychosocial support was integrated as part of the Health section of Project SPHERE in 2004 (Standard 3: mental and social aspects of health). The sections acknowledge the importance of psychological and sociological impact on the mental health of survivors. We were to propose language that was included two guidance notes in the guidance: (1) psychological first aid and (2) community based psychological interventions.

In 2006–2007 the Interagency Standing Committee (IASC) began conversation in preparing guidance for mental health and psychosocial support. Dr. Prewitt Diaz represented American Red Cross in the deliberations. The results of which was the current guidance for undertaking field based programs.

Psychosocial support is currently a cross cutting theme in the 2011 SPHERE project. Most of the topics generated in the early Red Cross sponsored programs are today included as guidance in the current documents. We have gone a long way, there is much needed in terms of mass casualty events and the incidence of mass migrations because of conflict in several parts of the world.

REFERENCES

IASC. *IASC Guidelines on Mental Health and Psychosocial support in Emergency settings.* Inter-Agency Standing Committee: Geneva, Switzerland, 2007.

NDMA. *National Disaster Management Guidelines: Psychosocial Support and Mental Health Services in Disasters.* National Disaster Management Authority (NDMA): New Delhi, 2009.

Prewitt Diaz, J. O.; Saballos, M. *Salud Psicosocial en un desastre complejo: El efecto del Huracan Mitch en Nicaragua.* Cruz Roja Nicaragüense y Universidad para la Paz: Managua, Nicaragua, 2000.

Prewitt Díaz, J. O.; González Flores, B. A. *Guia Metodologica: Diplomado para Interventores en Crisis.* Cruz Roja Guatemalteca: Ciudad de Guatemala, 2002.

Prewitt Díaz, J. O. *Apoyo psicosocial en desastres: Un modelo para Guatemala.* Cruz Roja Guatemalteca: Cuidad de Guatemala, 2002.

Prewitt Díaz, J. O.; Srinivasa Murthy, R.; Lasksminarayana, R. *Disaster Mental Health in India.* Indian Red Cross Society: New Delhi, India, 2004.

SPHERE Project. *Humanitarian Charter and Minimum Standards in Disaster Response.* The SPHERE Project: Geneva, Switzerland, 2004.

SPHERE Project. *Humanitarian Charter and Minimum Standards in Humanitarian Response.* Geneva, Switzerland; 2011

PART I

Theoretical Building Blocks and Background

CHAPTER 1

THEORETICAL BUILDING BLOCKS FOR COMMUNITY-BASED PSYCHOSOCIAL SUPPORT

JOSEPH O. PREWITT DIAZ[1], S.P. AGARWAL[2]

[1]*Chief Executive Officer, Center for Psychosocial Support Solutions, Alexandria, VA 22310, USA*

[2]*Secretary General of the Indian Red Cross Society 1 Red Cross Rd. New Delhi 10001, India*

CONTENTS

1.1 INTRODUCTION

Disasters, either natural events or human generated, impact and the emotional fabric of the survivors. The physical wounds can be seen, treated, and eventually be healed with appropriate treatment in a short period. The emotional wounds cannot be seen, and because they go unnoticed they linger on in the psyche and the memories of the affected people.

As a consequence of a disaster the minds and bodies of the affected people enter an unprecedented state of heightened psychological, social, and neurological arousal (Dahlitz, 2015). The affected people are focused on their immediate needs, regardless of preexisting relationships with the community. High brain arousal activates survival instincts. The brain focuses on the threat and on survival, so the body releases physical and psychological resources to deal with immediate survival needs (Sherin and Nemeroff, 2011). The effort is such that the worldview becomes narrow and so the attention to self, neighborhood, and community awareness is reduced. Important matters for affected individuals are to replace the perception of the physical, environmental, and social loss with intense impressions that dominate the experience of hurricane survival. The disaster-affected people undergo a radical process of psychosocial reorganization in order to deal with the present. This set of events lead to psychosocial detachment that are exclusive of the perception of this new reality which exclude the future and the past.

This process of psychosocial detachment results in a profound interruption of preexisting continuity and continued development of the psychosocial aspects of life in their communities. Since detachment is unfamiliar, the affected people cannot recognize the psychosocial interruption in their being, or understand its impact on their future psychosocial growth and evolution in place. This psychosocial detachment affects the normally constant, taken for granted, and not consciously experienced continuity of life (Gordon, 2004). There are three factors that impact the continuity of life: (1) the subjective sense of imminent death, injury, and helplessness; (2) how immersed the affected people were in the feeling, sounds and fury of the winds, rain, and wave of destruction; and (3) the duration of the perceived threat and identifying when safety and security could be reestablished.

If the human body is expected to be lost in death, it may lead to bodily disconnections (psychosomatic symptoms), lack of interest in sensations (food, warmth, or sexuality), numbness, disassociation, and out of body experiences as a result of traditional beliefs. These symptoms are associated with traumatic stressors. The emotional toll on the survivor's psychosocial detachment has major implications for the future:

- How many of the friends, family members, and known community members died, disappeared or were left for dead, or whose fate is unknown?
- Where are the relatives, friends, and community members who were separated through migration or involuntary movement within the country as a result of the hurricane?
- How have the individual lives of the affected people changed after facing death and now demonstrate a loss of motivation, enjoyment of daily interactions with family and friends, and other normal community activities if their occupation, homes, neighborhoods, and possessions are no longer considered important?
- Future ambitions, goals, and purpose of life have been terminated by the hurricane.
- In many individuals, there is a loss in the sense of place. Preparation for death means losing one's self, which causes identity problems and survival guilt, or it is changed by the hurricane.

Disasters cause complex consequences in the bodies, minds, and social systems of the affected population. Detachment and loss of the sense of place affect the continuity of social relationships. The psychosocial effects of the disaster rupture and/or degrade the physical, environmental, and figurative sense of place that supports the affected people and sustains place attachments to each other.

After a disaster, and in spite of the slow process of recovery, the affected people that remained in their neighborhoods and villages have begun to reestablish their place by forming a new survival oriented communal system in which the high state of detachment has shifted into a high energy reestablishment of place stage. The affected people are developing a commonality in regards to the disaster experience and their loss of place. Now, the communities have gained a collective significance for each other and for the larger society.

The well-being and recovery of affected people depend on a dynamic process of interactions between the affected people's needs and the resources available. To determine the needs of the affected, the psychosocial support personnel conducts on-site interviews with survivors, explored historical images, and investigated disaster area maps and photos of the destruction (Prewitt Diaz, 2013).

The holistic approach of mobilization takes into consideration the results of an assessment that considered the needs and resources of the affected location over time and assisted the affected people to identify the geographical, psychological, and social capital available to them. In addition, this assessment identified interventions to maintain individual's well-being. A deeper appreciation of the living, place-based experiences of the affected people would enable the stakeholders to consider more ethical and sustainable forms of community change.

Sharing stories and experiences from the past (i.e., Has this happened here before? How did your elders cope with the situation?) will revive lessons in the affected people that can evolve into creative practices in reestablishing place. The affected people will think differently about the new possibilities in at least three ways:

First, the inclusion of all affected people creates the feeling of being visible to partners and external stakeholders. Through their narratives and performances in community theaters, drawing or writing, the affected people can document their presence (how they used, moved through, and made their place, neighborhood, and community). The affected people can assert the right to be a part of the recovery (healing process) of their place, neighborhood, and community.

Second, they communicate and enact their experiences of place as inhabited; an understanding based on psychic attachments, bodily and social memories as well as fragile social ecologies. Affected people, as they consider the tensions of acceptance of the loss of everyday life in their place, are encouraged through community-based psychosocial support (CBPSS) to remember life in an otherwise previous part of an affected neighborhood and community. One potential difficulty is that in thinking what they have been through, there are many neighbors that have left the community and they will have to remake their life without them. We must be aware that the grieving process will be magnified to include the human loss of neighbors and friends.

Third, if the affected people are understood as having been wounded by the recovery activities, or lack thereof, and the devastating power of the disaster itself, other images of place might focus attention on why places, people, groups, environments, and nonhuman nature continue to be injured. If the affected people, their place, and neighborhood are wounded through displacement, material devastation, and root shock, then so is the whole community and its inhabitants, too.

CBPSS may be one way to assist the affected people in sharing their history, visualizing their experiences, and through community walks beginning to transform loss into reestablishment of place, a new beginning, and feelings of resilience. Psychosocial support may offer possibilities of place-based mourning and care across generations that build self-worth, collective security, and social capacity. Materially, CBPSS activities motivate the creation of social capital, provide a range of memorialization activities, create new forms of public memory, and are committed to intergenerational education and social outreach (Prewitt Diaz and Dayal, 2008). On the other hand, people who stayed in their places and communities have begun to involve themselves with others in the evacuation, registered for the Federal Emergency Management Agency (FEMA) and those receiving aid from different sources, and identified activities that strengthen the community fabric. The personal experiences have been put aside to focus on the reestablishment of their place.

Places may act as thresholds through which the living can contact those who have gone before, those who have migrated, and with human and nonhuman lives born and yet to come. If one takes root shock seriously, then caring for a place is a way to repair our worlds that may help create socially sustainable cities across generations. Through psychosocial activities, the affected people are called upon to be resident experts, to perceive the disaster destroyed geographical location as an inhabited place. In this matter, slowly, slowly, the affected people begin to unfold and open pathways of memory that generate possibilities of shared belongings in a reestablished place. The affected people, through psychosocial support activities, accept responsibility for being unprepared for a disaster, engage in mourning activities, begin a process of emotional transformation, and visualize and develop a new place as a group. Addressing the loss becomes a dynamic path of development, where we can discard the ruins caused by disasters and embrace a new beginning, that is, a reestablished place.

Several factors interact in the growth and welfare of an affected person. These factors are geographical (space, ecological factors), physical (food, shelter, protection, medical care), psychological, (attachment, affection, self-esteem), spiritual (belief system, identity, and values), and social (family, friends, sense of place to which one belongs). People belong and interact in each time and space with the community where they find themselves, whether it is the original or an adopted community. Therefore, to increase recovery and reconstruction and enhance resilience and well-being, we should encourage the affected people to act on various levels: individual, family, place, and community.

1.2   PSYCHOSOCIAL SUPPORT

An operational definition of psychosocial support is shared with the reader upfront so that we may understand the strategy used to reestablish a sense of psychological and social well-being. Psychosocial support reestablishes the social and psychological well-being of the people and provides the tools to rebuild their social networks. This approach provides the tools and spaces so that people recognize the emotional sequelae of the disaster, express their feelings, and initiate a process of reconstructing their lives within the social networks in the neighborhood and community. One frequently used intervention, that has been deemed evidence-informed, is psychological first aid (Ursano et al., 2007).

Psychosocial support activities are initiated immediately in the aftermath of a disaster by members of the family or the surviving community. These activities help to compensate for the sorrow and stress with an abundance of personal warmth and direct help. As in the Haiti earthquake, the community represented informal mass social and physical support: people were rescued, sheltered, and reassured by their community members.

Generating or reinforcing community organization projects, informal support networks as well as facilitating processes of creative problem solving and the development or strengthening of community projects. The affected people assess their community needs identify projects, undertake the project and monitoring, and evaluate the results. By initiating the project cycle psychosocial support activities provide a space for achieving some economic recovery.

Psychosocial support recognizes that the affected people themselves are the subjects of their own recovery by regaining the knowledge of their personal emotional needs, community environment, and the steps that should be taken for reconstruction. It is an integral, interdisciplinary, interinstitutional, and intersectoral process (Prewitt Diaz, 2008) that provides the tools to individuals, families, and the community to reestablish their resiliency and their social, functional, and psychological development; in such a way that they can resume or recreate their life.

1.3   COMMUNITY-BASED PSYCHOSOCIAL SUPPORT

The approach has two predominant segments: a clinical and a psychosocial segment. The clinical approach begins with psychological first aid, and then the small groups of disaster-affected people that need further clinical evaluation are referred for counseling. In a small number of cases, medical personnel that may use medication to alleviate the traumatic stress, anxiety, and depression arising from the difficulty of coping with the disasters and the feelings of hopelessness and helplessness experienced by some (WHO, 2011).

The psychosocial support segment provides planning, development, and monitoring for participatory community assessment. The community is an active participant in its own recovery (Prewitt Diaz, 2013). Most of this work is qualitative and relies on community mapping and narratives using pictures as stimuli for affected community members to express their needs. Psychosocial support is synonymous with constant activities, increased communication, ownership for the process and product, voluntarism, and reestablishment of place.

CBPSS is a compassionate tool that provides support and equal access to services to the oppressed and the oppressors. Psychosocial support is participatory in nature and provides a mechanism where all segments of the community can identify the risk and resilience factors in their geographic, ecological, cultural, economic, spiritual, social, and psychological place (De Jong, 2011). The activities provide a physical and temporal space for all members of a community to identify their losses, what they need to rebuild, their social capital, and what the affected-communities need from the outsiders and other stakeholders.

A CBPSS addresses basic needs (food, shelter, protection, and health) that provide safety, security, and increased comfort (Hobfoll, 2007). It promotes self-esteem and a sense of belonging to a place. Celebrate traditional, religious, and cultural aspects that promote growth in their place. Value and use the resources of the person, the place, and the community. The program fosters the reconstruction of children and adult their lives, family, group, and social fabric. Increase social access through the school, place, and community.

1.4   CONCEPT OF OPERATION

Effective CBPSS abides by a "rights-based ethic of care." These practices are grounded in stories, cultural practices, and natural beauty of the place. They are fundamental in the establishment of differentiated and active forms of belonging and just and equitable reestablishment of a place (Till, 2012). Place caring as an activity and a set of psychosocial relations should be at the heart of CBPSS activities (Till, p. 10).

Volunteers are the backbone of community-based psychosocial support. They will develop safe spaces for children to play, and informal schooling for adolescents, women, and the elderly to develop skills to rebuild their "place."

Slowly, slowly, the disaster-affected people regain their desire to move ahead and enhance their resilience. The narratives with small groups every 6 months give a glimpse of the movement made by the affected communities from their own perspectives and identify evolving needs for moving forward to continue, grow, and reestablish themselves.

Following the continuum of recovery, meaning should be achieved. Meaning is operationalized as the immediate consciousness of life prior to the impact of a disaster. Lived experience exists in its representation in the affected person; it does not exist out of memory. The relationship between memory and the lived experience is at the center of knowledge production in gaining an understanding of the meaning attributed to places by the affected people. A way to understand the meaning that affected people's attribute to a place is through sharing of their stories.

It is through social interactions that place meaning, derived from memories of the lived experience, are represented to a broader audience through community mobilization activities (Schwartz, 1989). As the affected people share the stories of their experiences and what it was like to be in place prior to the disaster, the people are constructing memories and sharing them with neighbors and other stakeholders.

Condensation, elaboration, and invention are common elements of remembering. The form in which people condense or streamline their memories and stories is constantly in flux, that is, they are impacted by the external social environment (Wertsch, 1998). People choose what they remember and how they represent those memories. In this way, people attribute meaning to their stories about their pre-disaster experience.

At a community level, affected people intertwine personal stories into a public (community) memory. Sharing stories in group meetings focus the dialogue by mediating multiple perspectives and interpretations.

Understanding place meaning improves the dialogue between affected persons and other stakeholders by making the explicit lived experience and the subsequent dialogue. While different members of the affected communities may have different levels of literacy, education, and community responsibilities, they all have experiential knowledge. When community meetings provide a space for individuals to share memories and stories of their experiences, they lead to important insights into place meaning. Conducting community "learning circles" allows for emotional knowledge to become crystallized into the collective meaning of a given disaster, what you have lost and where you are going. This community social interaction provides a conduit to identify place meaning that is derived from memories and lived experience, to be shared with a broader audience.

A disaster causes emotional transformation on the affected people (Hobfoll, 2007). To understand place meaning is to understand the emotional transformation of space to place. The emotional energy while sharing stories will lead the community to explore and develop strategies to move forward in reestablishing place. In a community, the participants of a meeting may experience two types of emotions: feelings of the experiences and the feelings of sharing them with others in a public forum.

Feelings of the lived experiences immediately associate the affected person with their environment in ways that they can clearly be understood by the audience.

The role of psychosocial support programs in its initial stages, irrespective of where the stakeholders believe they are in the disaster cycle, is seeking place meaning, is described as a way of constructing history itself, of inventing it, and providing different versions of "what happened here" based on community members' stories. For every identified "place meaning" and when these constructions are accepted by other people as credible and convincing, they enrich the common stock on which everyone can draw to understand past events, interpret their significance, and imagine how these events will happen anew. In the final analysis, the nature of place meanings or sense of place is predicated by the changing experiences, feelings, and memories of the affected people. How the affected people feel about their neighbors, families, and communities, and places of the past and present, will shape their memories and their stories into an emerging new place.

1.5   THE PHYSIOLOGY, GEOGRAPHY, AND PSYCHOLOGY OF PLACE

Psychosocial support is the basis for the reestablishing of place. This section presents the literature that supports the theoretical framework for this essay. Some of the early literature describes place from a physiological perspective. Hofer (2002) explains the growth of a seed in these terms; a healthy seed cannot grow into a plant without fertile soil, light, and water, we resist to recognize the importance of the environment in our life. Better yet, how place molds our behavior and optic when facing a problem such as a major destructive disaster.

Hofer explains the role of place in the most basic terms. A Place is important for human beings right from the beginning. For example, sperm and egg are influenced by their immediate surroundings. If an egg is not in the right place it cannot form the placenta. The same principle applies when a zygote is only a few cells as when a whole baby is interacting with its world. Nature and nurture cooperate in producing behavior, the organism has a lot to say about what it is going to get out of its environment. A good bit of that say depends on the organism being in the right place at the right time.

1.5.1 PLACE

Home is a place that holds great importance to disaster-affected people. Home can be a material place or an imagined place. Home is an encompassing category that combines place/physical location of the physical structure of a house with an emotional set of meanings like permanence, continuity, and rootedness (Blunt and Dowling, 2006). "Home" is not only a three-dimensional structure, a shelter, but it is also a matrix of social relations and has wide symbolic and ideological meanings, a place of unconditional love where we are accepted for who we are. Home can reflect a place of one's own or a place of shared belonging, and where we feel emotional links to a greater community (Robinson and Adams, 2008).

This section presents a discussion on the six relationships associated with a feeling of "place." To understand the behavior changes experienced by disaster-affected persons after a major disaster, humanitarian workers need to understand physical, social, psychological, and spiritual changes experienced by the survivors. The six relationships are (1) history, memories, and experiences, (2) subjective and objective self, (3) rules that inform living in a place, (4) family and community stories, common events, vegetation, culture, music, and spiritual experiences, (5) choosing the place where the person wants to be, and (6) circumstantial and opportunistic relationships.

1.5.2 HISTORY, MEMORIES, AND EXPERIENCE

The strongest and most enduring relationships are based on the length of residence. The time spent interacting within a place create memories and experience that form the individual and community identity.

At the heart of the person's psychological structure is a sense of belonging. Behavior and feelings of "place belongingness" are necessary and the basis for place identity. Place identity is built on the social, cultural, spiritual, and biological definition of cognitions of place that becomes a part of a person's place identity (Dixon and Durkheim, 2000).

The interaction of the physical, psychological, social, and spiritual self with geographic and geographic and environmental "places" has been studied. Jacobs (1961) and Gans (1968) explored community social dynamics and how they enrich planning practices. Relph (1976) studied place and a sense of place, and Proshansky (1978) looked at the city and self-identity. McMillan (1976) attempted to define the variables that comprised "sense of community." Appleyard (1979) studied people's perception and attitude toward the place and began to explore place meaning to inform the planning process. These researchers concluded community focused emotions, cognitions and behaviors can impact community planning and development (Manzo and Perkins, 2006, p. 336)

1.5.3 SUBJECTIVE AND OBJECTIVE SELF

The interaction between self, others, and the environment is difficult to measure. The connections to place are sometimes intuitive and mystical. Disaster-affected people also experience losses to their subjective "self" that can only be understood by consultation and involvement of disaster-affected people in a description about the nature of their subjective losses, and what steps must be taken to reconstruct "self within the place." Or better said, reconstructing the interaction between the disaster-affected person and the environment that they identify as "their place."

A balance between subjective and an objective self-guide the way human beings understand, accept, and behave in their "place" (Coppersmith, 1968). Subjective self includes those innermost feelings about family members, communities, and self, and is predicated by our attachments mechanism within as people grow and mature. The interaction of an individual with the environment and self generates what has been identified as self-esteem. Self-esteem is a deep set of values and believes that control a person's behaviors and attachment (Coppersmith, 1968).

Hummon (1992) defines a sense of place as a subjective perception and feeling that people have of their environment. The sense of place has a dual nature; on the one hand it interprets the environment, and on the other there is an emotional reaction to the environment. Hutton indicates that "sense of place" involves a personal orientation toward a place in which persons' understanding of the place and the persons' feelings about the place become fused in the context of environmental meaning.

Hummon (2002) describes "place" as a notion that is concerned with location, their meanings, and the ways people relate to them. Place attachment reflects the involvement of individuals within their social, psychological, spiritual, and environmental geographical areas. Community satisfaction is directly related to place attachment by predictors such as friendliness of the people, personal safety, physical environment, spiritual meaning. Kelly and Hosking (2008) found positive relationships between levels of place attachment and the amount of time spent in the geographical area, support of local economic enterprises, and contributions through voluntary organizations.

1.5.4   RULES ABOUT HOW TO LIVE IN A PLACE

At times disaster-affected people are resettled for long periods of time in trailer parks or in communities far away from where they originally lived (i.e., FEMA trailers cities after Hurricane Katrina). Disaster-affected people should develop a set of rules that dictate how they will behave in this "place." The government or other outside group (i.e., religious community) set the rules for all community members and their actions.

1.6   FAMILY AND COMMUNITY STORIES, COMMON EVENTS, VEGETATION, CULTURE, MUSIC, AND SPIRITUAL EXPERIENCES

Disaster-affected people grow up hearing stories about "place," whether real or mythical. Everyone has the same language, play in the same places, has the same trees, and other parts of the environment is always in a persons' sight.

At the heart of the affected person's psychological structure is a sense of belonging and place identity (social, cultural, spiritual, and the biological) that compose a person's place identity (Dixon and Durkheim, 2000). Proshansky et al. (1983) describe place identity as a "potpourri of memories, conceptions, interpretations, ideas, and related feelings about specific physical settings as well as types of settings (p. 60)."

The physical sense is expressed by a tacit knowledge of physical details of place, the psychosocial details, recognizes the connection with local communities by recognizing their integration. Our identification with "place" is a location where affected people keep personal belongings, memory books, diaries and photographs of our past life, and a central gathering place for family and friends. The degree of identification we develop with "our place" can be so powerful that after a disaster, one of the first things affected people feel impelled to do is to return to "my place."

The innermost feeling of belonging to a place (family, neighborhood, or community) are shattered and the community networks that traditionally back up the affected people are no longer available. After a major catastrophe, affected people lose their environment, tangible geographic areas, and infrastructure. Those things are easy to "eyeball" they are part of the objective self: how the affected people dressed, behaved, and interacted with the community and environment after the disaster or emergency. The objective self includes three senses (physical, social, and autobiographical) (Rowles, 1983).

The objective self (recognized as a persons' self-concept) is overt, how we dress, behave, interact with the community and the environment around us. The objective self includes three senses: physical, social, and autobiographical (Rowles, 1983). The physical sense is expressed as a kind of tacit knowledge of physical details of place; the social self recognizes a connection with the local community by recognizing their integration within the social fabric, and the autobiographical sense that arises from a person's transactions within place over time.

After a disaster has forced the affected people from their physical surroundings and has damaged the environment beyond recognition, people can maintain a sense of belonging by remembering incidents, places, contributions, and relations from their personal lives in the place.

This dichotomy between the subjective and objective determines our formulation of self in interaction with community and environment. Self becomes in places. Places are conceived as dynamic arenas that are both socially constituted and are constitutive of the social (Dixon and Durkheim, 2000) which elicit an individual and collective identity. Proshansky et al. (1983) describe place identity as a "potpourri of memories, conceptions, interpretations, ideas, and related feelings about specific physical settings as well as types of settings" (p. 60).

Korpela (1989) defines place identity as a psychological structure that arises out of an individual attempt to regulate the environment. Fishermen who survived the tsunami now say that they want to return to the sea. They identify their experience in the sea as "love," "freedom," or "challenge." The sense of place is a construct that we create within ourselves in time (Jackson, 1994). It is the result of habit or custom, and it is reinforced by a set of recurring events.

Glynn (1981) identified two dimensions of a community; territorial and relational. The relational dimension of a community had to do with the nature and quality of relationships in the community. Some communities may even have no discernable territorial demarcation, as in the case of a community of carpenters working in a specialty who have contact and quality relationships, but may live and work in different locations, perhaps even throughout the western coast of Sri Lanka. Other communities are defined primarily per territory, as in the case of neighborhoods, but even in such cases, proximity or shared territory cannot by itself constitute a community; the relational dimension is essential.

The dependency factors between the survivor and the environment form a dynamic process by which the disaster-affected people manipulate their environment to satisfy their psychological, social, and spiritual needs. The sense of place denotes a dynamic process that a disaster-affected person manipulates the environment to reflect their identity, thus giving an emotional meaning to the person-environment relationship after a disaster (Lazarus, 1966; Lazarus and Folkman, 1984).

1.7 CHOOSING A PLACE WHERE A PERSON WANTS TO BE

Choice of a place may dictate where a person moves, too. As urbanization occurs in places around big cities, people tend to displace themselves into the suburbs looking for quiet, greener, areas and the convenience of living of shopping and recreation. The decision to move to the new "place" is more cognitive than emotional. After a disaster, the affected people may be forced to move to a new place that has been chosen for them (i.e., FEMA trailers in a different city).

As individuals grow up there are several sources of attachment that allow them to maintain self-coherence, self-esteem, and self-regulation. The person becomes attached to their mothers for nurturance, they attach with family members for security, and they attach to the community for well-being. Attachment to place is a factor that people rely on to feel safe. Safety is a human need to be able to function in their world. A feeling of safety allows a person to feel in control of their abilities to handle difficult situations. Place implies support from others who can be relied on in time of difficulty. The sense of safety that affected people feel in "place" is based on the expectation that psychosocial networks will provide emotional support and comfort.

Williams and Stewart (1998) indicate that place is composed of meanings, beliefs, symbols, values, and feelings that individuals or groups associate with a specific locality; the emotional bonds that people form with places, at various geographic scales, over time; the strongly felt values, meanings, and symbols that are hard to identify or know; the valued qualities of a place that may not be consciously aware of until they are threatened or lost; the set of place meanings that are actively and continuously constructed and reconstructed within individual minds, shared cultures, and social practices; and the awareness of the cultural, historical, and spatial context within which meanings, values, and social interactions (Williams and Steward, 1998).

This relationship between self and environment are restricted to the choices available for the disaster-affected person. Usually the very young or the elderly are found amongst the people in this category. In the current economic condition people may move because they have found a job in the new place or the family may have gotten transferred from one military installation to another.

1.8   PLACE ATTACHMENT

Place attachment is related to the emotional ties that disaster-affected people have with their places (home and natural environment). This bond is maintained through the interaction with social, material, and physical environments (Charis and Lawrence, 2014). The intensity of people's place attachment can differ depending on the number of contact people have with a place, the size and location of the place, and whether the place is threatened. Studies have found that people report higher place attachment to their homes than to their neighborhoods (Hidalgo and Hernandez, 2001; Anton and Lawrence, 2014). This could be because the home is a more easily definable space with obvious boundaries, whereas "neighborhood" or "local area" is harder to define as they lack obvious boundaries or property lines.

Tanner (2012) indicates that attachment becomes part of a person's identity and can evoke positive feelings of trust, safety, belonging, and rootedness that encourage people to invest in a locality. Development of shared histories and experiences promotes the development of emotional safety and security for people that are part of the community.

Place attachment refers to "the emotional or affective bonds which a person feels to an area or place" (Livingston et al., 2008). In this book, the attachment is utilized for practical reasons. A place may have several differing boundaries; in this book, we are interested in attachment to the neighborhoods and communities where disaster-affected people live. Place attachment can take two forms after a disaster: functional (practical) attachment (i.e., widows or widowers, children, the elderly), and emotional attachment (i.e., "this is where I was born," "that is where I played dominoes with my friends," "that is where I kept my boat"). Affected people tend to form a stronger bond to place if it meets their physical and psychological needs and matches their lifestyles. People become attached to a place; (1) if place supports their lifestyle and self-identity, (2) place must be distinct from others, in the eyes of the affected people, (3) provide a continuity of experience over time, and enable to disaster-affected people to make a positive evaluation of themselves, thus increasing their self-esteem.

This definition of attachment is having a positive impact on both individuals and neighborhoods. It encourages participation in community activities in a positive way. Place attachment tends to be higher for older people and those that have lived in a defined geographical place for a long time. The longer a person lives in a community the higher the trust in the public safety and the neighborhood. Place attachment is associated with strong social networks, evidence of community cohesion, and low criminal activities.

Place attachment is often thought to be higher in more homogeneous communities where the inhabitants have common backgrounds, interests, cultural and religious affiliations and lifestyles (Twigger-Ross and Uzzzell, 1996). In the tsunami, it was easier to develop programs amongst a community of fisherman, Muslin or Christian communities or amongst the community women group than among heterogeneous groups.

Social cohesion and security in place are the two most important components of place attachment. The regeneration of disaster affected communities into sustainable communities is predicated on encouraging the develop strategies that will strengthen social cohesion, psychosocial networks, and physical and emotional well-being.

At the personal level, the longer a person lives in a place the more positive their sentiments towards the neighborhood and community are going to be (Korpela, 1989). The more education a person has in a place (i.e., the only schooling that I have done was right in this community school), the greater the attachment to the place.

In the community context, a person is more likely to feel attached to a place where others feel attached, where they have many friends and family members, and where they have greater involvement in local organizations (i.e., clubs, churches, sports teams). These groups foster a positive evaluation of self, the feeling of social value, and a sense of pride (Twigger-Ross and Uzzell, 1996).

To reconstruct the community networks after a disaster there are at least three actions that should be considered: first, identify the factors that are important for the development of place attachment by disaster-affected people; second, examine what impact pre-disaster social structures and social mix have on attachment, for example, in the Gujarat earthquake it was very difficult to bring together persons from different castes. This cast difference resulted in a shelter remaining open for over five years. While in Sri Lanka, groups belonging to the same religion, came together in a short period to reconstruct their housing areas; third, assess the extent to which place attachment activities may remain disaster-affected people in the same geographic areas for a long period. The problem being that instability may undermine the psychological, social, and spiritual networks and cohesion that are needed to build sustainable communities post-disaster.

Place attachment has a positive effect on disaster-affected people, helping to enrich their lives with meaning, values, and significance, thus contributing to people's mental health and well-being (Twigger-Ross and Uzzell, 1996). Functional attachment occurs when the ability of a place to enable a person to achieve their desired goals and activities.

1.9   SATISFACTION

Achieving satisfaction with place includes the interaction of three distinct factors. The first is the natural and built environment as well as the values and attitudes toward that environment. The second refers to the psychological connection and dependence on the place which is often separated into identity and dependence on the place. The third is connected to the social community as well as cultural context. The fourth dimension is composed of the behavioral reaction to the sociocultural environment.

The effect of the natural and built environment includes people's contact with their neighborhood and communities, knowledge of the plants and the trees, the green areas, and where to find shade, or fruits or the smell of flowers. What are the climatic conditions of place (for example, in my place I can expect that the morning will be cool, but in the afternoon when the sun shifts it is extremely hot, and then as soon as the sun goes down it is cool once again)? Another consideration is the geography (frequently people who live in the mountains will behave differently than those who live in the valley or near the sea).

The second factor is the psychology of the people. It is divided into an emotional and a functional attachment. The length of residence allows development of a self-concept and interactions with others in their neighborhoods or communities. Participation in community activities presents a person the opportunity to develop deep relationships with others and with the environment (Theodori and Luloft, 2000).

The third factor is the sociocultural factor; it suggests that connection to place is rooted in social interactions within the neighborhood or community. The sense of place derives from people, experiences, and memories that allow for sharing of stories, celebrations, and commemorations. The cultural component is related to the development and recognition of symbols that the community groups use to produce memories from the place (Hidalgo and Hernandez, 2001). Ardoin (2006) suggests that people connect to place through relationships that generate through lifestyle choice, familial history, a feeling of a sense of belonging.

1.10 MEANINGS

Meaning is operationalized as the immediate consciousness of life prior to the impact of a disaster. Lived experiences exist in their representation in the affected person; they do not exist out of memory. The relationship between memory and the lived experience is at the center of knowledge production in gaining an understanding of the meaning attributed to places by the affected people. To understand the meaning that affected people attribute to place is through sharing of their stories.

It is through social interactions that place meaning, derived from memories of the lived experience, are represented to a broader audience through community mobilization activities (Schwartz, 1989). As the affected people share the stories of their experiences and what it was like to be in place prior to the disaster, the people are constructing memories and sharing them with neighbors and other stakeholders.

Condensation, elaboration, and invention are common elements of remembering. The form in which people condense or streamline their memories and stories is constantly in flux, that is, they are impacted by the external social environment (Wertsch, 1998). People choose what they remember and how they represent those memories. In this way, people attribute meaning to their stories about their pre-disaster experience.

At a community level, affected people intertwine personal stories into a public (community) memory. Sharing stories in group meetings focuses the dialogue by mediating multiple perspectives and interpretations.

Understanding place meaning improves the dialogue between affected persons and other stakeholders by making the explicit lived experience and the subsequent dialogue. While different members of the affected communities may have different levels of literacy, education, and community responsibilities, they all have experiential knowledge. When community meetings provide a space for individuals to share memories and stories of their experiences, they lead to important insights into place meaning. Conducting community "learning circles" allows for emotional knowledge to become crystallized into the collective meaning of a given disaster, what you have lost and where you are going. This community social interaction provides a conduit to identify place meaning that is derived from memories and lived experience, to be shared with a broader audience.

A disaster causes emotional transformation on the affected people (Hobfoll, 2008). To understand place meaning is to understand the emotional transformation of space to place. The emotional energy while sharing stories will lead the community to explore and develop strategies to move forward in reestablishing place. In a community, the participants of a meeting may experience two types of emotions: feelings of the experiences and the feelings of sharing them with others in a public forum. Feelings of the lived experiences immediately associate the affected person with their environment in ways that they can clearly be understood by the audience.

The role of psychosocial support programs in its initial stages, irrespective of where the stakeholders believe they are in the disaster cycle, in seeking place meaning is described as a way of constructing history itself, of inventing it and providing different versions of "what happened here" based on community members' stories. For every identified "place meaning" and when these constructs are accepted by other people as credible and convincing, they enrich the common stock on which everyone can draw to understand past events, interpret their significance, and imagine how these events will happen anew.

In the final analysis, the nature of place meanings or sense of place is predicated by the changing experiences, feelings, and memories of the affected people. How the affected people feel about their neighbors, families, and communities, and places of the past and present will shape their memories and their stories into an emerging new place.

1.11   PLACE-BASED SECURITY

Disaster-affected people feel secure when they have some control over their life. A sense of security involves a belief that one can protect their place from perceived danger. They can be psychosocial or physical threats. A sense of security can be the perceived safety and physical protection, or it can exist in the form psychosocial protection. The sense of security is a connection to place. Psychosocial security and freedom from anxiety and fear play an important role in a person's sense of security. Disaster-affected people have a need to feel secure and free from threads, they are aware that maintaining social relations serves as a deterrent to feeling lonely or isolated.

Attachment to a place can provide a feeling of security, belonging, and stability to affected people. A sense of security appears because of the relationship of people to places. Feelings of security offer psychosocial protection from the unknown, threats of anxiety, instability, loneliness, and discontinuity, lack of control uncertainty. The sense of security is the confidence in the routine and reliability of family and neighbors, places and things (Elliot et al., 2000).

Hawkins and Maurer (2011) identified more specific places like the home place and community as key to the development of security because in this environment, routines are promoted that develop a sense of order, safety, and trust in the world (Hawkins and Maurer, 2011). Dupuis and Thorns (1998) identify four elements in the home that assure security: (1) it is a site of constancy in a social and material environment, providing continuity and a sense of permanence; (2) it is a space in which daily routines and activities are performed, so a sense of predictability and familiarity is developed; (3) it is a place where people are most in control of their lives and have privacy, providing a place of refuge and autonomy, and (4) it provides a secure base from where identities can be constructed and developed.

1.12   LOSS OF PLACE AND RELOCATION

Disasters disrupt place attachment and lead to a loss of security (Hawkins and Maurer, 2011). They also disrupt the processes that bind people to their social and material environments causing a decline in place attachment. Chamlee-Wright and Storr (2009) found that the psychological and physical loss of home and community significantly influenced participants' ideas about safety, routine, and trust in a stable environment. This loss can lead to a feeling of anxiety, helplessness, hopeless and stress.

Probably the worst consequence of a disaster is the loss of place and relocation. The affected people are separated from their social, geographical, psychosocial, and economic environments and forced to move away to an unknown situation. Disasters result in forced relocation because of the damage done by natural events. Frequently, the involuntary relocation is sudden and can overwhelm the sense of safety and security. Although disaster response agencies assist in the reconstruction of homes and replacing material things, the emotional results of loss of place result in years of grief and a lodging to result in the place left behind. The place needs to be understood using a multidisciplinary approach.

The greatest need that a human being has is the ability to control one's experience. Whether we are happy or sad, or happy or depressed is related to the place where we decide to spend our life. Our actions, thoughts, and feelings are shaped not only by our physical or psychological makeup, but by our place.

There are three fundamental aspects of place: location, locale, and sense of place (Fullilove, 1999). Location refers to the site (Cresswell, 2004), the material components that give shape to the place where people live their lives make up the locale (Cresswell, 2004). The sense of place is the attachment people should place that is both subjective and emotional (Creswell, 2009). The loss of that place, the emptiness generates a series of emotions: grief, disorientation, alienation, and dislocation, to name a few (Fullilove, 1999). The loss of place ruptures the emotional connections with place leaving a feeling of dislocation characterized by nostalgia, alienation, loss of attachments, and familiarity (Fullilove, 1976).

Fullilove (2004) calls the "root shock" the forced removal by natural or man-made causes, encountering the trauma of displacement throughout the disaster-affected area. Displacement root shock is "the traumatic stress reaction to the destruction of all or part of one's emotional ecosystem" (p. 11). Fullilove draws a parallel of root shock to the physiological shock experienced by an affected person who loses massive amounts of fluid because of injury, a shock that threatens the whole body's ability to function.

Fullilove describes the place as having a central function in an individual's emotional and social ecosystem. For Fullilove, the place is a kind of exoskeleton (personal conversation with author, 2007). As such, we can understand the place as always becoming, as within and beyond us, and as functioning as a kind of social protective shell.

Fullilove's (2004) notion of maze ways clarifies how embodied social spaces provide a personal and social shell; when affected people are forced to leave their neighborhoods and their protective maze ways are destroyed and/or cannot be made, root shock, per Fullilove, may stay with a person for a lifetime. Fullilove suggests that the physical fabric of the neighborhood one grows up in also provides the cues and opportunities for the intergenerational transmission of stories, and as such root shock may affect multiple generations. Thus, root shock can be inherited through psychological, social, bodily, and place memory.

Places are, thus, both personal and social, made of human and nonhuman lives. Through making and maintaining places, individuals sustain the maze ways and enacted assemblages created through personal, bodily, social, and material worlds (Fullilove, 2004). Ecological character-istics provide structure for individuals and social groups and may include places of worship, schools, marketplaces, and parks. Social institutions that are located throughout places and neighborhoods offer regular activi-ties and rhythms for local and social communities.

Taken together, these social ecologies of place include everyday routines, social institutions, material landscapes (the fabric, taste, sounds, and scents of places), symbolic systems of meaning and identity, and shared memories. When those ecologies are damaged, individuals or groups will most likely experience root shock (Fullilove, 2004).

Places become part of us, even when held in common, through the intimate relationships individuals and groups have with places. Such an understanding resonates with the ways that affected people speak about individual and shared places, as having a distinct presence, material, sensual, spiritual, and psychic values (Chamlee-Wright and Storr, 2009, pp. 622–623).

While these every day, embodied knowledge of inhabited places may be taken for granted, when residents are forcibly relocated as a result of a disaster, individuals may become aware of the intensity of their place-based attachments which have become severed, which may give rise to another level of loss that is deeper-seated and more chronic than the imme-diate losses resulting from destruction and/or displacement (Fullilove, 1976). When places are physically demolished not only is the stability of the taken-for-granted rhythms destroyed, an individual's personal and intra-subjective emotional ecosystems become damaged. At the same time, Fullilove notes that when individuals try to reconnect after root shock, their personal understanding of structural and institutional violence might result in the creation of healing places.

1.13   DYNAMIC INTERACTION OF AFFECTED PEOPLE AND RESOURCES IN PLACE

The welfare and recovery of the affected people depend upon a dynamic process of interactions between the affected people needs and resources in place. To determine the needs of the affected people the CBPSS personnel conduct on-site interviews with survivors, explore historical images, elicit maps of the place, and displays photos with scenes of destruction.

The holistic approach of mobilization takes into consideration the results of an assessment consider needs and resources of the affected place in time, and assist the affected people to identify the geographical, psychological, and social capital available to them, and at the same time, identify interventions for the well-being. A deeper appreciation of the living, place-based experiences of the affected people would enable the stakeholders to consider more ethical and sustainable forms of community change.

Sharing stories and experiences from the past (Has this happened here before? How did your elders cope with the situation?) will revive lessons in the affected people that will evolve into creative practices in reestablishing place. The affected people will think differently about the new possibilities in at least three ways. First, the inclusion of all affected creates the feeling of being visible to partners and external stakeholders. Through their narratives and performances in community theaters, affected people documented their presence (how they used, moved through, and made their place, neighborhood, and community). First, the affected people can assert the right to be a part of the recovery (healing process) of their place, neighborhood, and community. Second, they communicated and enacted their experiences of place as inhabited, an understanding based on psychic attachments, bodily and social memories, and fragile social ecologies. Affected people, as they consider the tensions of acceptance of the loss of everyday life in their place, are encouraged through CBPSS to remember life in an otherwise previously part of an affected neighborhood and community.

If the affected people are understood as having been wounded by the disaster risk reduction activities, or lack thereof, and the devastating power of the disaster itself, other images of place might focus attention on why places, peoples, groups, environments, and nonhuman natures continue to be injured. The active metaphor of "wounded place" is operational in the psychosocial realm (Till, 2012). If the affected people, their place, and neighborhood are wounded through displacement, material devastation, and root shock, so, too, is the whole community and its inhabitants.

Disaster wounds uncover old wounds (Till, 2012), be those tied to the histories of the affected people in the place, processes, and traumas of displacement. The unwrapping of old wounds and common stories about the place are important in terms of intergenerational relations and silences, and the affected individual, families and neighborhood capacity to repair. Community-based psychosocial support may offer possibilities of place-based mourning and care across generations that build self-worth, collective security, and social capacity (Till, 2008). Materially, CBPSS activities motivate the creation of social capital, provide a range of memorialization activities, creates new forms of public memory, and is committed to intergenerational education and social outreach (Prewitt Diaz and Dayal, 2008).

1.14   THE COMMUNITY-BASED MODEL (CBPSS)

The program that will be the subject of this chapter was called "Community-Based Psychosocial Support" (CBPSS). The objective of the program was to provide a space to reestablish a sense of place after a disaster, or the attachment that disaster-affected people have toward their homes, neighborhoods, and communities after having experienced displacement and loss (Prewitt Diaz, 2013).

There are three components to the program: (1) developing a safe environment for the group of affected people to begin to process their loss and develop a plan, (2) learn helpful behaviors to navigate through the safe space, (3) develop a realistic plan with members of their group to move on to reestablish place, neighborhood, and community. The project initiates community engagement with those who show a readiness to engage in reestablishing place. The development process that is identified and initiated by disaster-affected people in their communities represents the whole community at any given time.

1.15   DEVELOPING A SAFE ENVIRONMENT

Safe environments were discovered by the therapeutic community movement in 1940. Bion developed a program whose purpose was to shift the "figure-ground" distribution of power. He thought that using all the relationships and activities in a safe space would aid the therapeutic task (Bion, 1946). In his study, he identified "place" as a small group based on expressed interests; the task of this group was to assist its members to manipulate and harmonize the personal relationships, reduce intergroup tension, and develop a feeling of safety and security among its members.

CBPSS explores the internal tension of place in the aftermath of a disaster. This program is inclusive of gender, age, diverse cultures and castes, handicapping conditions, and migratory patterns. Each small group, safe space, or community is representative of this diversity.

Ring et al. (2005) developed a program that helped the affected people to learn, manipulate, and harmonize personal relationship in a place to achieve feelings of safety and security. A community place is formed by many personal places (disaster-affected people feel safe and secure: home, locality, and locale). The objective is to let disaster-affected people identify intragroup tensions and develop some ideas about harmonizing those tensions. They identify what ecological modifications must take place so that the community feels holistic calmness within. This guided activity is geared to have the disaster-affected people, see for themselves what has taken place in their microcosm, and identify what are the recovery tasks (What happened, how do we resolve the situation, where do we get the resources, what will our place look like after we fix it).

The affected people are encouraged to identify a "life space." A "life space may be the spot where two paths intersect," the soccer field or a quiet place under a tree. This is a place within the community accessible to all the members of the community that may be used for meetings with external stakeholders, to hold discuss among community members and reduce tension, and moments of large gatherings such as community cooking, religious ceremonies, concert, or plays. It is in these "life spaces" that the disaster-affected people identify the social capital within the community, what additional external resources are needed and what the external stakeholders may provide.

1.16 LEARNING HELPFUL BEHAVIORS

Helpful behaviors are those that facilitate the engagement of disaster-affected people into small group processes. The CBPSS is an inclusive program and begins its intervention by providing a space for affected people to understand themselves as related to what they have just survived, and the impact that verbal and nonverbal behavior have in their place. This approach uses psychological first aid as a tool for group and individual learning focusing on inclusiveness and protection. The group begins to identify helpful behaviors, inclusive behaviors, and behaviors that impede group growth.

The facilitator encourages to develop their autonomy by (1) conducting orderly interventions and listening to others; (2) problem focused inter-action (we/they relationships), and (3) creative problem-solving using a solution-focused approach. The methodology was an adaptation from National Training Laboratories (NTL, 1987). The expected outcomes are: (1) to develop leadership styles that may help the participants reach out to the affected community; (2) understand the group process; (3) develop self-awareness and learn how the participants are perceived by others; and (4) transition from a perception of victim to victors, as community psycho-social facilitators exhibit the skills of change agent (Campbell et al., 2003).

After completing this training phase, the facilitators in training increase self-awareness of behavior and interpersonal style in relation to others, learn to relate and communicate more effectively with others, develop abilities to give and receive feedback effectively, and can develop commu-nity (NTL, 1999).

1.17   DEVELOP A REALISTIC PLAN THAT ACHIEVES SATISFACTION

Achieving satisfaction with place includes the interaction of three distinct factors. The first is the natural and built environment, as well as the values and attitudes toward that environment. The second refers to the psycholog-ical connection to and dependence on the place which is often separated into identity and dependence on the place. The third is connected to the social community as well as cultural context.

The first factor is the effect of the natural and built environment includes people's contact with their neighborhood and communities, knowledge of the plants and the trees, the green areas, and where to find shade, or fruits or the smell of flowers. What are the climatic conditions of place? (e.g., in my place I can expect that the morning will be cool, but in the afternoon when the sun shifts it is extremely hot, and then as the sun goes down it is cool once again) Another consideration is the geography (frequently people that live in the mountains will behave differently than those that live in the valley or near the sea).

The second factor is the psychology of the people. It is divided into emotional and functional attachment. The length of residence allows development of a self-concept and interactions with others in their neighborhoods or communities. Participation in community activities gives a person the opportunity to develop deep relationships with others and with the environment (Theodori and Luloft, 2000).

The third factor is the sociocultural factor that suggests that connection to place is rooted in social interactions within the neighborhood or community. The sense of place derives from people, experiences, and memories that allow for sharing of stories, celebrations, and commemorations. The cultural component is related to the development and recognition of symbols that the community groups use to produce memories from the place (Hidalgo and Hernandez, 2001). Ardoin (2006) suggests that people connect to place through relationships that generate through lifestyle choice, familial history, and a feeling of a sense of belonging.

1.18   SUMMARY

This chapter has introduced the theoretical building blocks for a community-based program focusing on psychosocial support. It covered place, attachment, and root shock. Overall, if the disaster planners are trying to measure activities used to reestablish place the most important learning to carry forward is that place and sense of place are qualitative concepts. The meanings and values that affected people give to place evolve over time. The place can have a different meaning to different people, for example, after a disaster, such as a tornado, the affected population will have a different perception than those who were not affected. The concept of place evolves over time and, thus different members may differ over the meaning of place; for example, children, adolescent, and elderly may express meanings that are discrepant, and so the selection of community attributes may be different for each group. Finally, sense of place may be an elusive concept to study since its limit is difficult to determine.

REFERENCES

Anton, C. E.; Lawrence, C. Home is Where the Heart is: The Effect of Place of Residence on Place Attachment and Community Participation. *J. Environ. Psychol.* **2014,** *40*, 451–461.

Ardoin, N. M. Sense of Place and Environmentally Responsible Behavior: What the Research Says. 2006.

Barton, A. *Communities in Disasters*; Basic Books: New York, NY, 1969.

Berry, J. W.; Kim, U.; Boski, P. (1988). Psychological Acculturation of Immigrants. In *Cross-Cultural Adaptation: Current Approaches;* Kim, Y. Y., Gudykunst, W. B., Eds.; Sage: Thousand Oaks, CA.

Blunt, A.; Dowling, R. (2006). *Home*. Routledge: London.

Brown, B. B.; Perkins, D. D. Disruption in Place Attachment. In *Place Attachment:* 1992; Altman, I., Low, S. M., Eds.; Plenum Press: London, England.

Chamlee-Wright, E.; Henry-Storr, V. There's no Place like New Orleans: Sense of Place and Community Recovery in the Ninth Ward after Hurricane Katrina. *J. Urban Affairs* **2009,** *31*(5), 615–634.

Dahlitz, M. J. Neuropsychotherapy: Defining the Emerging Paradigm of Neurobiologically Informed Psychotherapy. *Int. J. Neuropsychotherapy* **2015,** *3*(1), 47–69. DOI: 10.12744/ijnpt.2015.0047-0069.

De Jong, K. *Psychosocial and Mental Health Interventions in Areas of Mass Violence,* 2011 2nd ed.; Doctors without Borders: Amsterdam.

Dupuis, A.; Thorns, D. C. Home, Home Ownership and the Search for Ontological Security. *Social. Rev.* **1998,** *46*, 24–47.

Elliott P.; Wakefield, J.; Best, N. G.; Briggs, D. J. *Spatial Epidemiology: Methods and Applications*. CAB International: Oxfordshire, UK, 2000.

Fullilove, M. T. (1976). Psychiatric implications of displacement: contributions from the psychology of place. American Journal of Psychatry. https://doi.org/10.1176/ajp.153.12.1516

Fullilove, M. T. *Root Shock: How Tearing Up City Neighborhoods Hurt America, and What We Can Do About it.* One World/Ballentine Press: New York, 2004.

Giuliani, M. V. Theory of Attachment and Place Attachment. In *Psychological Theories for Environmental Issues;* Bonnes, M., Lee, T., Bonaito, M., Eds.; Ashgate Publishers: Aldershot, 2003.

Gordon, R. The Social System as Site of Disaster Impact and Resource for Recovery. *Aust. J. Emerg. Manage.* **2004,** *19*(4), 16–22.

Hawkins, R. L.; Maurer, K. You Fix My Community; You have Fixed My Life: The Disruption and Rebuilding of Ontological Security in New Orleans. *Disaster: J. Policy Manage.* **2011,** *35*(1), 143–159.

Hidalgo, M. C.; Hernandez, B. Place Attachment: Conceptual and Empirical Questions. *J. Environ. Psychol.* **2001,** *21*, 273–281.

Hobfoll, S. E.; Watson, P.; Carl C.; Bell, C. C.; Bryant, R. A.; et al. Five Essential Elements of Immediate and Mid-Term Mass Trauma Intervention: Empirical Evidence. *Psychiatry* **2007,** *70*(4), 283–315.

Hofer, M. A. The Riddle of Development. In *Conceptions of Development*; Lewkowicz, D. J., Lickliter, R., Eds.; Psychology Press: Philadelphia, 2002; pp 5–29.

Inter-Agency Standing Committee (IASC). *A Common Monitoring and Evaluation Framework for Mental Health and Psychosocial Support in Emergencies*. IASC, Reference Group for Mental Health and Psychosocial Support in Emergency Settings: Geneva, 2017.

Kar, N. Psychological Impact of Disasters on Children: Review of Assessment and Interventions. *World J. Pediatr.* **2009**, *5*(1), 5–11.

Korpela, K. M. Place Identity as a Product of Environmental Self-Regulation. *J. Environ. Psychol.* **1989**, *9*, 241–256.

Livingston, M.; Bailey, N.; Kearns, A. *The Influence of Neighborhood Deprivation on People's Attachment to Places*. Communication Department, Joseph Roundtree Foundation: York, England, 2008.

MHPSS. *Mental Health and Psychosocial Support in Humanitarian Emergencies: What Should Humanitarian Health Actors Should Know?* IASC Reference Group for Mental Health and Psychosocial Support in Emergency Settings. Interagency Standing Committee: Geneva, Switzerland, 2010.

Perry, R. W.; Lindell, M. K. The Psychological Consequences of Natural Disaster: A Review of Research in American Communities. *Mass Emergency.* **1978**, *3*, 105–115.

Prewitt Diaz, J. O. Integrating Psychosocial Programs in Multi-Sector Responses to International Disasters. *Am. Psychol.* **2008**, *63*(8), 818–827.

Prewitt Diaz, J. O. Community-Based Psychosocial Support: An Overview. In *Disaster Management: Medical Preparedness, Response and Homeland Security;* Arora, R., Arora, P., Eds.; CAB International: Oxfordshire, UK, 2013.

Prewitt Diaz, J. O.; Dayal, A. Sense of Place: A Model for Community Based Psychosocial Support Programs. *Australas. J. Disaster* **2008**, *1*, 1–15.

Project SPHERE. *Humanitarian Charter and Minimum Standards in Humanitarian Response*. Project SPHERE: Geneva, Switzerland, 2007.

Robinson, E.; Adams, R. Housing Stress and the Mental Health and Well-being of Families. Australian Family Relationships Clearinghouse Briefing, 12. 2008, pp 1–9. (accessed Jan 5, 2016).

Schwartz, D. Visual Ethnography: Using Photographs in Qualitative Research. *Qual. Sociol.* **1989**, *12*(2), 119–153.

Sherin, J. E.; Nemeroff, C. B. Post-Traumatic Stress Disorder: The Neurobiological Impact of Psychological Trauma. *Dialogues Clin. Neurosci.* **2011**, *13*(3), 263–278.

Stedman, R. Is it Just a Social Construction? The Contribution of the Physical Environment to Sense of Place. *Soc. Nat. Resources.* **2003**, *16*, 671–685.

Tanner, K. Place Attachment and Place-based Security: The Experiences of Red and Green Zone Residents in Post-earthquake Kaiapoi. Doctoral Dissertation, University of Canterbury, New Zealand, 2012. http://www.geog.canterbury.ac.nz/postgrad/420papers/2012/Kimberley_Tanner_420_dissertation.pdf (accessed Jan 5, 2016).

Theodori, G. L.; Luloff, A. E. Urbanization and Community Attachment in Rural Areas. *Soc. Nat. Resour.* **2000**, *13*, 399–420.

Till, K. E. Wounded Cities: Memory-Work and a Place Based Ethics of Care. *Political Geography* **2012**, *31*(2012), 3–14.

Tronto, J. C. *Moral Boundaries: A Political Argument for an Ethic of Care*. Chapman & Hall, Routledge: New York, 1999.

Twigger-Ross, C. L.; Ussell, D. L. Place and Identity Processes. *J. Environ. Psychol.* **1996,** *16*, 205–220.

Ursano, R. J.; Fullerton, C.; Benedek, D. M.; Hamaoka, D. A. Hurricane Katrina: Disasters Teach Us and We Must Learn. *Acad. Psychiatry* **2007,** *31*(3), 180–182.

Uzzell, D.; Pol, E.; Badenas, D. Place Identification, Social Cohesion and Environmental Sustainability. *Environ. Behav.* **2002,** *34*(1), 26–53.

Wallace, A. Mazeway Disintegration. *Hum. Organ.* **1957,** *16*, 23–27.

Wertsch, J. V. *Mind and Action*. Oxford University Press: New York, NY, 1998.

Williams, D. R.; Vaske, J. J. The Measure of Place Attachment: Validity and Generalizability of a Psychometric Approach. *For. Sci.* **2003,** *49*(6), 830–840.

World Health Organization (WHO). *Psychological First Aid: Guide for Field Workers*. World Health Organization, World Trauma Foundation & World Vision: Geneva, Switzerland, 2011.

CHAPTER 2

HISTORICAL OVERVIEW OF RECENT POLICY STATEMENTS, GUIDANCE, AND AGREEMENTS PERTAINING TO MENTAL HEALTH AND PSYCHOSOCIAL SUPPORT

JOSEPH O. PREWITT DIAZ

Chief Executive Officer of the Center for Psychosocial Support Solution, Alexandria, VA 22310, USA

CONTENTS

2.1   INTRODUCTION

The increase in natural disasters since 2000, and the introduction of disaster mental health, trauma counseling, and psychosocial support as modalities used to alleviate suffering and promote comfort on disaster-affected people. Scholars and practitioners to indentified the most appropriate methodologies to apply after a disaster that would cause "no harm" to the recipient. The Sphere Project (2004) acknowledged that there were mental health issues related to the emotional response to a disaster. The guidance identified psychosocial support as a tool to alleviate fear and promote resilience in the affected communities, thus recognizing that disaster-affected people may experience immense psychosocial suffering. The guidance also recommends mental health services from professionals who have previous mental health issues that are exacerbated by the disaster experience. This was an early attempt to protect the right of people with mental or intellectual disabilities. The guidance suggests that loss of place may affect the recovery process and a timely achievement of well-being.

2.2 HISTORICAL EVOLUTION

The United Nations (UN) Hyogo picked up and commented on the role of psychosocial support in disasters. The UN Hyogo Framework or Action 2005 (UN/HFA, 2005), under the rubric of "Social and Economic Practices," the guidance indicates that we should "enhance recovery schemes including psychosocial training programs to mitigate the psychosocial damage of vulnerable (disaster-affected people) particularly children, in the aftermath of disasters" (Paragraph 19 (g)). The UN undertook some major efforts in the following years.

In 2007, after 2 years of deliberation, consultation, and field testing, the Inter-Agency Standing Committee (IASC) published guidelines on mental and psychosocial support in emergencies. The guidelines consist of a single document published in 2007, comprising a set of key principles as well as a matrix of key interventions spanning emergency preparedness, minimum responses, and a comprehensive response. The key interventions include coordination, human resources, community mobilization, health, nutrition, water, and sanitation (Mayer and Loughry, 2014). The guidelines use the term "mental health" and "psychosocial support" to describe any type of local or outside intervention that aims to protect or promote psychosocial well-being or to prevent or treat mental disorders. Although both terms overlap in the humanitarian world, they reflect different approaches (Van Ommeren and Wessells, 2007).

In 2010, a group of experts invited by the World Health Organization (WHO) to discuss the integration of mental health and psychosocial support (MHPSS) in the Millennium Development Goals (MDGs) (WHO, 2010) concluded the following:

1. Mental health issues should be integrated into all broader development and poverty eradication policies and programs as a key indicator of human development.

2. Education programs should integrate mental health and psycho-social perspectives in efforts to improve quality of education and ensure accessibility for people with mental and psychosocial disabilities and intellectual disabilities.

3. Mental health and psychosocial issues should be integrated into all efforts to promote gender equality and empowerment of women including efforts to fight against gender-based violence.

4. Sustainable development requires that policies and programs related to climate change, disaster reduction and response, and slum and urbanization integrate the mental health and psychosocial perspective.

5. Cooperation efforts must ensure the participation of people with disabilities including those with mental and psychosocial disabilities. Efforts to improve information and communication technologies need to integrate the mental health and disability perspectives (WHO, September 2010).

These recommendations lead the reader to understand that mental and psychosocial needs lead to poverty, high unemployment or underemployment, and poor educational and health outcomes. Moving forward, there is a need for a wider recognition of mental and emotional well-being as a core indicator of human development. Furthermore, it is necessary to integrate a mental and psychosocial perspective into all humanitarian policies, as well as programs and services (WHO, 2010, p. 10).

In 2013, the UN expert group meeting at Kuala Lumpur was held to discuss psychological well-being, disability, and development. Among its key goals, the group identified the role of the international community to take all appropriate measures to ensure that the human rights of persons with mental or intellectual disabilities are protected and promoted, in line with the Convention on the Rights of Persons with Disabilities, in gender- and age-sensitive manners with a life-cycle approach, with special attention to women, children, adolescents, older persons, and other marginalized populations such as migrants, refugees and internally displaced persons, and indigenous people. The group concluded that mental well-being (i.e., psychosocial support and maintenance of a sense of place) should be integrated into all social development efforts. Protection and promotion of the rights of people with the mental or intellectual disorder should be integrated and strengthened as a key priority in the future discussion (Conclusions of the UN Expert Group meeting at Kuala Lumpur, 29 April–1 May 2013).

In 2014, the UN expert group on mental well-being, disability, and disaster risk reduction (DRR) met in Tokyo, Japan. The group highlighted several consequences of disasters in mental health and psychosocial well-being. Among those consequences are: (1) Disaster-affected populations frequently experience immense mental and psychosocial suffering. Although most people are capable of handling with life's challenges, MHPSS need to be made available for those who require it in support of their recovery. (2) Protection and promotion of mental and psychosocial well-being along with the rights of persons with mental or intellectual disabilities tend to face further challenges. For some affected populations, disasters can trigger psychological symptoms or exacerbate existing mental health conditions. Though many people can handle with psychological distress after disasters without professional help, some require nonspecialized or specialized interventions. (3) These conditions can have long-term consequences, medically, psychologically, socially, and economically and can affect recovery and reconstruction if not addressed. The impact of poor mental well-being is pervasive and can lead to morbidity and mortality, low productivity, social unrest, poverty, inequality, high unemployment, and delays in recovery and reconstruction (UN Expert Group, pp. 4–6).

In their deliberation, the group of experts recommended adding targets and indicators on mental health and psychosocial well-being in DRR. The HFA 2 should include mental health and psychosocial well-being as transformative new targets and as indicators to represent subjective well-being toward optimizing resilience of people and society. Some of the actions suggested by the group were: (1) raise awareness among all people, including decision-makers and regarding the importance of mental and psychosocial well-being; (2) facilitate recovery from psychosocial distress to regain and advance productivity. Improve social support and the mental health system in the preparedness, response, recovery, and reconstruction phases; (3) provide information on how to protect and promote mental health and psychosocial well-being and optimize resilience, including self-care and in risk information (UN expert group, p. 13).

The Sendai Framework for DRR 2015–2030 under priority 4, Enhancing disaster preparedness for effective response and to "Build Back Better" in recovery, rehabilitation, and reconstruction, talks about "empowering women and persons with disabilities to publicly lead and promote gender equitable and universally accessible response, recovery, rehabilitation, and reconstruction approaches is key" (p. 20). The priority point further suggests that "disasters have demonstrated that the recovery, rehabilitation, and reconstruction phase, which needs to be prepared ahead of a disaster, is a critical opportunity to 'Build Back Better,' including through integrating DRR into development measures, making nations and communities resilient to disasters." It concludes that "to enhance recovery schemes to provide psychosocial support and mental health services for all people in need" (p. 21).

In 2007, the IASC guidelines on MHPSS in emergency settings were launched. The stated purpose of the guidelines was to offer advice on how to facilitate an integrated approach to address mental health and psychosocial issues in emergency situations. Inter-Agency Standing Committee/ Mental Health and Psychosocial Support (IASC/MHPSS) manual identifies the primary objective of these guidelines is to enable humanitarian actors and communities to plan, establish, and coordinate a set of minimum multisectoral responses to protect and improve people's mental health and psychosocial well-being during an emergency (p. 5).

To complement the focus on minimum response, the guidelines also list concrete strategies for MHPSS to be considered mainly before and after the acute emergency phase. These "before" (emergency preparedness) and "after" (comprehensive response) steps establish a context for the minimum response and emphasize that the minimum response is only the starting point for more comprehensive supports (IASC/MHPSS, 2007, p. 6). The guidelines introduce a function matrix that is divided into three parts. The first part is entitled "common functions" (coordination, assessment, monitoring, evaluation, and human resources). The second part is entitled "core mental health and psychosocial functions," these include: community mobilization, health services, education, and dissemination of information. The third part is called "social considerations in sectorial domain," and it includes: nutrition, shelter, and water and sanitation (p. 20).

Psychosocial support has been included as a crosscutting issue in the Project Sphere 2011 (Project Sphere, 2011, p. 17). The main role of psychosocial support is to address the complex emotional, social, physical, and spiritual effects of disasters. It strengthens the humanitarian action by engaging the affected people in guiding and implementing the disaster response, by promoting self-help, coping, and resilience (p. 17).

In 2013, the IASC Reference Group on MHPSS introduced an assessment guide. This guide has been used for rapid assessments of MHPSS issues in recent humanitarian emergencies. The guide introduces three tools that are designed to build on three types of information: (a) existing information collected through desk reviews, (b) collecting new information through assessments by MHPSS actors, and (c) collecting new information through integrating MHPSS questions in assessments by different sectors (p. 3). The guide introduces three templates, one for each type of information.

In 2014, the IASC/MHPSS ordered a review of the implantation of the guidelines after 7 years of use in the field. The review looked at the level of implementation of the guidelines and supplementary tools, and the mainstreaming and integration of the guidelines across the humanitarian system since the guidelines were first published. The results were based on key informant interviews, extensive document review, online survey, and in-depth and brief case studies (p. 7).

The review of implementation includes a review of academic literature, including articles relevant to the key principles in the guidelines, as well as research that has directly referred to or utilized the guidelines as a conceptual framework or basis for research. More specifically it explores the overall influence of the guidelines, as described by key informants, awareness, utilization, and institutionalization of the guidelines (IASC, 2013, p. 13).

Findings of the review indicate that the impact of the guidelines has been widespread and significant. Seven years of dissemination, utilization, and implementation of the guidelines in a range of vastly differing contexts has offered an opportunity for reflection, consolidation, and mapping of the next steps toward strengthening and improving MHPSS response in emergencies. The findings demonstrate that the impact of the guidelines has been widespread and significant. The levels of awareness of the guidelines are mixed, depending on the context; however, overall knowledge of the existence of the guidelines is high, while deeper knowledge of the content may require additional efforts. The guidelines have been used to improve program quality, yet further guidance on psychosocial programming may be useful. The influence of the guidelines on programs and activities in the field appears to be positive, however, this influence depends on resources, context, and capacity (IASC 2013, p. 55).

The Sphere Project began to address MHPSS as a noncommunicable disease under the health guidance since 2000. It is not until 2004 that the Sphere guidance acknowledges psychosocial support and includes Standard 3, under the minimum standards in health Sphere, 2004, pp. 291–293.

The explanation of the standard reads as follows: "Social and psychological indicators are discussed separately. The term 'social intervention' is used for those activities that primarily aim to have social effects. The term 'psychological intervention' is used for interventions that primarily aim to have a psychological effect (p. 291)."

The guidance notes provide the reader with some new information. The manual suggests that information (a way of reducing public anxiety and distress) should be provided. The most important information is (1) the nature and scale of the event, and (2) efforts to establish physical safety, types of relief, and who is offering the same and location to assure security and calming (p. 292), (3) a cadre of respondents should be trained in psychological first aid to assist the disaster-affected population as well as staff. Psychological first aid (PFA) is composed of nonintrusive pragmatic care with focus on listening, assessing needs, and assuring that basic needs are met, and promoting company from significant others (p. 293). (4) Community-based psychological interventions should be based on an assessment of existing services and the sociocultural context.

These interventions should include use of functional, cultural coping mechanisms for individuals and communities to help them gain their control. Collaboration with community leaders and local healers is encouraged, as well as self-help groups. Community workers should be trained and supervised to assist with heavy caseloads or to conduct outreach activities and care to vulnerable and minority populations in the affected areas (p. 293).

The 2001 Gujarat earthquake, the 2005 Southeast Asia tsunami, and the 2005 Hurricane Katrina brought humanitarian actors to the table. The Sphere project in the 2011 edition shared the impact off the deliberations by humanitarian partners. Their objective was to revise the Sphere project to reflect gaps in service during these large disasters. Key to these deliberations were the principle of "Do no Harm." Two basic actions were identified: (1) what were common rights and duties of humanitarian workers and (2) the need to be accountable to the disaster-affected communities.

The Humanitarian Charter and Minimum Standards in Disaster Response (2011) revealed the conclusions reached by these deliberations among humanitarian agencies should "ensure that their actions do not bring further harm to the affected people" (Sphere Project, 2011, p. 6). The activities should benefit the most affected and vulnerable, and that the partners contribute to protect disaster-affected people from violence and other human rights abuses. Thus, one of the crosscutting themes was psychosocial support: "Some of the greatest sources of vulnerability and suffering in disasters arise from the complex emotional, social, physical, and spiritual effects of disasters.

Many of these reactions are normal and can be overcome with time. It is essential to organize locally appropriate MHPSSs that promote self-help, coping, and resilience among affected people. Humanitarian action is strengthened if at the earliest appropriate moment, affected people are engaged in guiding and implementing the disaster response. In each humanitarian sector, the way aid is administered has a psychosocial impact that may either support or cause harm to affected people. Aid should be delivered in a compassionate manner that promotes dignity, enables self-efficacy through meaningful participation, respects the importance of religious and cultural practices, and strengthens the ability of community people to support holistic well-being" (p. 17).

In 2017, the Reference Group for MHPSS in emergency settings published a common framework for monitoring and evaluation of psycho-social support programs. This product was by a group of scholars and practitioners underfunding of UNICEF and The Office of U.S. Foreign Disaster Assistance/Office of U.S. Foreign Disaster Assistance (OFDA/ USAID). It provides a common set of guidance chapter that assist the person in the field with identifying program goals and objectives, devel-oping a Log Frame, and provides practical tips for a common framework, and gives a schematic for sharing results and lessons learned. The value of this clear and well written document is that in the near future the prac-titioners will be able to measure a common set of actions under the rubric of MHPSS.

As this book goes into print the 2nd draft for consultation of The Sphere Handbook for humanitarian response has been posted on the Internet   (http://www.sphereproject.org/handbook/revision-sphere-hand-book/draft-ready-for-feedback/). The final guidance is due to be published in 2018. The basic difference in the 2018 Handbook proposes: (1) a new structure for the minimum standards; (2) changes in the context and ways of delivering humanitarian response, and (3) simpler language. The new handbook acknowledges differences across populations which makes the humanitarian response more inclusive.

In terms of MHPSS, the handbook considers the needs of the affected populations and acknowledges that MHPSS response requires compli-mentary support and coordinated responses across different sectors. Social and cultural considerations in provision of basic services and security are essential to reduce stress. The Handbook Introduction (2017) processes strengthening community psychosocial support and self help as conduits to a more protective environment that allows disaster-affected people to help each other to achieve social and emotional recovery (p. 12)

2.3   SUMMARY

Psychosocial support has evolved as the need for the protection of the most vulnerable after humanitarian crises, and the response have identified gaps in assuring that the recipient of services would not be further harmed by the interventions. The UN, the coalition of faith-based institutions, and the Red Cross internationally have brought forth the need and have taken on the identification of a remedy. Psychosocial support began as a social adjunct to the medical needs of disaster-affected people. Soon, thereafter the UN expert panels attached psychosocial support as part of the common services offered to people with special needs or disaster-affected people.

In 2007, IASC formed a group of experts from public and private sectors to discuss the problem and propose solutions. The chapter introduces these deliberations from this group and the proposed guidelines. The chapter concludes with a discussion of the evolution of psychosocial support as presented by the Sphere project. This development is important since it is the document that is used by humanitarian respondents after a disaster. Psychosocial support has become an important component of all international guidance as it is the sector that protects disaster-affected people from increased harm because of an immediate response to the disaster.

REFERENCES

IASC RG MHPSS. *IASC Reference Group Mental Health and Psychosocial Support Assessment Guide*; Inter-Agency Standing Committee: Geneva, 2017.

Inter-Agency Standing Committee (IASC). *IASC Guidelines on Mental Health and Psychosocial Support in Emergency Settings;* IASC: Geneva, 2007.

Inter-Agency Standing Committee (IASC). *Reference Group on Mental Health and Psychosocial Support in Emergency Settings*; Inter-Agency Standing Committee: Geneva, 2013.

Mayer, S.; Loughry, M. *Review of the Implementation of the IASC Guidelines on Mental Health and Psychosocial Support in Emergency Settings*; UNICEF Headquarters and the Reference Group: New York, November 2014.

Sphere Project. *Humanitarian Charter and Minimum Standards in Disaster Response*; The SPHERE Project: Geneva, Switzerland, 2004.

Sphere Project. *The SPHERE Project;* The SPHERE project: Geneva, Switzerland, 2011.

Sphere Project. *The SPHERE Handbook for Humanitarian Crisis (2nd draft) for Consultation.* 2017. (http://www.sphereproject.org/handbook/revision-sphere-handbook/draft-ready-for-feedback/ (accessed Dec 2, 2017).

UN. *The SENDAI Framework for Disaster Risk Reduction 2015–2030.* Sendai, Japan. Adopted by the Third UN Conference in Sendai, Japan, March 18, 2015.

UN Expert Group. *UN Expert Group Meeting on Mental Well-being, Disability, and Development*; Kuala Lumpur, Malaysia. United Nations, April 29–May 1, 2013.

UN Expert Group. *United Nations Expert Group Meeting on Mental Well-being, Disability, and Disaster Risk Reduction*; United Nations University Headquarters: Tokyo, Japan, Nov 27–28, 2014.

UN/ISDR. *HYOGO Framework for Action 2005–2015: Building the Resilience of Nations and Communities to Disasters*; World Conference on Disaster Reduction: Kobe, Hyogo, Japan, 2005.

Van Ommeren, M.; Wessells, M. *Inter-Agency Agreement on Mental Health and Psychosocial Support in Emergency Settings*; Bulletin of the World Health Organization: Geneva, Switzerland, November 2007; 85(11), p 822.

World Health Organization (WHO) *Mental Health and Development: Integrating Mental Health into all Development Efforts Including MDG's*; WHO. UN(DESA)-WHO Policy Analysis: Geneva, Switzerland. September 12, 2010.

CHAPTER 3

COMPONENTS OF COMMUNITY-BASED PSYCHOSOCIAL SUPPORT IN SRI LANKA: PLANNING, IMPLEMENTING, MONITORING, AND EVALUATING COMMUNITY RESILIENCE PROJECT

JOSEPH O. PREWITT DIAZ[1] and ANJANA DAYAL De PREWITT[2]

[1]*Chief Executive Officer, Center for Psychosocial Support Solutions, Alexandria, VA 22310, USA*

[2]*Associate Director Psychosocial Support Programs, Save the Children USA, Washington, D.C., USA*

CONTENTS

3.1 INTRODUCTION

This chapter focuses on the program management component of a psycho-social support program. It will use the Sri Lanka Red Cross Society/ American Red Cross (SLRC/ARC) Psychosocial Support Program (PSP) implemented post-tsunami to illustrate project management cycle. It is coauthored by the PSP technical advisor and the project manager of the SLRCS/ARC PSP.

Sudden onset of a disaster, such as the tsunami, taxes the capacity of the individuals and communities to make choices regarding their psycho-logical and social well-being. The traditional values, norms, and behaviors fail the survivor and, thus, the sense of powerlessness and "loss of place" is reported by many of the disaster survivors. Survivors' capacity to use the remaining resources in the community that can support their recovery and enhance their well-being is substantially reduced.

Although all survivors have the similar needs, the factors that meet those needs depend on and vary per the culture and context. Surviving a disaster is a human condition that promotes the presence of resilience factors in the form of psychosocial behavioral strategies and patterns. These factors and their perceived values have varying impact on the psychosocial well-being of the community of survivors. Understanding the community as a psychosocial entity permits the planners of long-term recovery programs to assist the community in identifying the factors that enhance resilience and promote well-being. The community can reestab-lish their sense of place through its own resources (Fullilove 1996).

The transition from the acute phase to recovery to reconstruction requires different methodologies with different objectives. During the acute phase, the survivors are provided with assistance to meet the basic needs (food, water, shelter, and clothing and, in few cases, emotional support). Many agencies and good Samaritans came to the aid of survi-vors by providing items indiscriminately as a form of charity. During the recovery and reconstruction phase, the attempt is to assist the survivors to enhance their resilience, recognize community resources, and identify ways to obtain sustainable livelihood. The way that agencies can help during these latter stages is through empowering the survivors and their communities to reestablish their "sense of place." The more the community takes charge of identifying their needs, existing resources, seeking local solutions, and engaging in problem-solving activities, the quicker they will recover, outgrow the effects of the disaster, and move on with their lives.

3.2   NATURE OF A COMMUNITY

Communities are dynamic and have a personality of their own. A community lives and functions constantly even though its human members come, go, born, or die. Community is a cultural system characterized by its diversity. Every community has different groups and points of struggle. Some of the most common groups that the community facilitator will encounter are based upon differences in gender, religion, access to wealth, ethnicity, class, educational level, income, ownership of capital, and language.

The concept of community may be defined (1) geographically (a group of people living near the shore or a municipality), (2) by common ethnic, cultural, religious, or linguistic characteristics, (3) by common ownership (fishing boat), (4) by a common occupation (the community of fisherman), by their families, (5) by common goals defined and agreed on. In the case of tsunami-affected areas, all communities and their ecological entities are exposed to the same destructive force.

A community is composed of a changing set of relationships, attitudes, and behaviors of its members. A human settlement is not merely a collection of houses; it is a social and cultural entity, too. Community-based PSP promotes community participation and self-help activities in a community of survivors. This means that the social organization of the community is changed slightly to increase social cohesion and inclusion. To achieve psychosocial well-being, the Red Cross volunteers should consider the interconnections between the cultural and power dimensions that comprise a community during the process of organizing the community.

3.3   REESTABLISHMENT OF SENSE OF PLACE

After a disaster, community's social fabric is destroyed due to deaths, destruction, and displacement. The survivors experience a root shock as they are uprooted from their day-to-day life routines and homes. Sri Lanka Red Cross Society/American Red Cross Psychosocial Support Program worked toward reestablishing the "sense of place" of the survivors to facilitate their recovery. However, this process needs to be defined and collaboratively carried out by the survivors themselves with some facilitation of the external actors.

In this case, the Red Cross staff reached out the community leaders, both formal and informal, to begin the process by asking questions and collating information through key informant interviews, focused groups, and workshops with other stakeholders utilizing tools such as free listing, community mapping, and community action plans. All the plans were to facilitate the reestablishment of sense of place and, thus improve the well-being and resilience of the survivors. The plans covered the five elements of sense of place, namely ecological, cultural, human capital, solution focused, and problem-solving activities (Fig. 3.1).

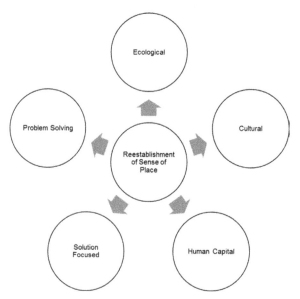

FIGURE 3.1   Five elements of sense of place.

3.3.1   CULTURAL

Cultures are the essence of what keeps the community alive. Just like communities, cultures have a life of their own which transcends its members. Culture is composed of language, symbols, the meaning that humans attribute to those symbols, and the overt and internal manifestations of behavior.

Cultures can live beyond the lives of community members. A basic definition of culture is a cumulus of individual's world view. People relied on what they knew, and yet many did not survive the force of the tsunami.

Community resilience projects or the reestablishment of place are a form of empowerment of the culture in a community. Community changes require modifications in cultural patterns to become stronger and survive. We can suggest methodologies and even instruments that may enhance well-being, but the community is the entity that makes the choice about modifications they need to make in their culture to move ahead and reconstruct.

To reestablish their "place," the survivors are encouraged to maintain their identity, and they are also asked to examine the way they think and the way they act to reconstruct their lives. Cultural activities that lead to reestablishment of "place" require the survivors to actively engage in making their own communal decisions, take time and try to choose their goals, identify resources and make their own community action plans to empower themselves and their communities.

3.3.2 ECOLOGICAL

Natural disaster often destroys the ecological environment of the impacted place. The tsunami that occurred in Sri Lanka destroyed the trees, plants, and even ecosystem some parts of the ocean with debris. In the floor bed of the ocean much debris had accumulated. Thus, some fishermen communities could not fish in their designated areas because there were no fish and it led to conflicts with other fishermen communities. This instance clearly illustrates the direct connection with ecology and psychosocial well-being. The first step consist of helping people to didentify the ecological segments of the community that were destroyed. The community members utilized the community's knowledge of native plants and wildlife; exploring and explaining the interaction between individuals and the environment; analyzing the comfort level between individuals and the environment; and then bringing those ecological segments back into the community.

3.3.3 IDENTIFYING HUMAN CAPITAL

A resource is something that can be potentially used as an input, as something to produce some other desired output. The community may not have money but they have human resources which may be considered the community contribution to a project. People help in the planning, monitoring, decision-making, management, report writing; these are the resources needed and can be provided from within the community.

Potential resources from the community can be useful, such as, a retired carpenter who may be willing to train some young members of the community, some unemployed youth who can provide energy and enthusiasm, some farmers or other food producers who are willing to prepare food for communal laborers who donate their time and energy, some loyal and trustworthy community members willing to put in time and thinking to design a community project.

Community resources include many noncash goods and services such as land, a place to locate the project, tools, raw materials, labor to provide human energy, as well as mechanical energy such as electricity, sun, wind, and water power. All the resources are identified and build upon to help the survivors reestablish their sense of place in terms of the activities they did before the disaster or would like to engage in going forward.

3.3.4 PROBLEM SOLVING

When the community is ready to design projects that enhance their well-being and build their resilience, the Red Cross and community volunteers help them identify and prioritize the problems they want to tackle. This process involves techniques that by engaging all the community members in the identification of the problems that affect their psychosocial well-being. Therefore, the community members need to learn the principles and skills of participation in evaluation, assessment, and appraisal. There must be an agreement and consensus among community members that the chosen problem to be solved is the one with the highest priority.

Once the problems are identified and prioritized based on severity and frequency, the community explores the resources within itself first to solve the problems. For example, in case of destroyed furniture, a retired carpenter of the community may train other members to manufacture new furniture for their families and might even use the new skill to generate income for their families.

3.3.5   SOLUTION-FOCUSED ACTIVITIES

Often after a disaster, individuals, families, and communities spend a lot of time on what is lost and the challenges that arise as a result of the loss. However, beginning to focus on solutions rather than mere challenges and problems, helps the survivors to move toward recovery and well-being. Red Cross staff helped the community members to look for solutions right from the assessment phase. They look at the problems but also look for available options to resolve those problems. For example, a group of senior students decided to paint the damaged furniture with the paint provided by Red Cross. Their activity was simple, enjoyable, and helped the students to make a meaningful contribution to their society.

CASE STUDY 1:UTILIZING COMMUNITY RESOURCES IN A RATMALANA INTERNALLY DISPLACED PERSONS (IDP) CAMP, SRI LANKA

The internally displaced persons (IDP) camp in Ratmalana was occupied by 37 families. There lived 157 men, women, and children together in a small compound. They had each other and very few worldly possessions. In identifying the needs of the community, they indicated that they needed beds. "Please get us off the floor." All they would need were 37 double beds. The Red Cross worker inquired further, "where we can get these beds?" he wondered out loud in front of the neighbors. One old man who was standing in the back finally answered the question. If you get me the wood and some of these young boys to help me, I will teach them how to make the beds. Six men began to work, measuring away and making calculations of the wood they would need, the cost of transport, and the design they would use to make the beds. American Red Cross staff went with a trusted community volunteer to purchase the wood and some tools. The next day, the carpenter and the volunteers began their task. Seven days later the members of the community had completed making all 37 beds. The community had identified its human resources and used them to attend to a community need. The good news made everybody happy and a community meal with activities for children and a skit by the community women helped to celebrate this small yet significant step ahead.

3.4   REESTABLISHING SENSE OF PLACE: PROJECT
MANAGEMENT

Community-based PSPs require flexibility because they involve the whole community and a variety of opinions. However, for these to be successful and reach their desired impact within the given time, these need to have concrete project management components. SLRCS/ARC (IASC, 2007) planning support system (PSS) utilized all the components of the project cycle, that is, participatory assessment, collaborative planning, implementation, monitoring, and evaluation. Since the project was designed with bottom-up approach, there was a need to build capacity of the staff and volunteers right from the community level.

3.4.1   CAPACITY BUILDING OF THE STAFF

Since community-based psychosocial support is a new field, it is usually difficult to find trained psychosocial support personnel in any context. However, the vast need of survivors' post-tsunami required immediate action. To do no harm, the leadership decided to provide a 2-day operational training for all the staff and volunteers who were going to interact with the survivors. The training included topics such psychological first, community assessment, self-care, and resilience. Once the program began to develop, the capacity-building continuum continued with several tiers of trainings such as PSS facilitation skills in schools and communities, crisis intervention technicians, specialists, and professionals. All the trainings included follow-up sessions and supervision by the experts in the field.

Capacity-building activities were not limited to the Red Cross staff but extended to the community leaders, school teachers, government officials, social workers, and faith-based leaders. The purpose was to enhance the skills of community members to understand their responsibility toward the psychosocial well-being of the community. The goal of the community is to be better prepared to plan, respond, and survive with their existing resources since it always takes some time before the help arrives from external sources (Dayal, 2006).

3.4.2 PARTICIPATORY ASSESSMENT

The project conducted a rapid assessment to identify and meet immediate needs of the survivors. Thus, project staff helped build temporary class-rooms which provided psychological first aid to children and adults who exhibited higher levels of stress reactions and also made referrals for other resources where relevant. However, once the project was fully staffed, an in-depth assessment was conducted.

Participatory assessment is one of the most frequently used tech-niques in assessing the community, determining risks, and identifying the community resources. One of the goals of participatory assessment is to magnify the voice of those community groups that are traditionally not heard. In the subsequent section, three forms of participatory assessment are discussed: (1) key informant interviews, (2) focused groups, and (3) community mapping.

3.4.2.1 KEY INFORMANT INTERVIEWS

SLRCS/ARC PSS staff identified entry points in the target communities and started the assessment process by conducting interviews with them. These were semi-structured interviews that included introduction and goal of the project, gathering ideas and suggestions of the interviewee and, where relevant, seeking permission to work in their community. Partici-pants were identified with consultation from the local Red Cross branch and with other community members.

Selection criteria included knowledge about the community, accept-ability, influence, and decision-making authority. The participants belonged to different religious, gender, and socioeconomic groups and, thus, provided a variety of perspectives. Key informant interviews were conducted throughout the project cycle to check satisfaction and engage-ment of people in the project.

3.4.2.2 FOCUSED GROUPS DISCUSSIONS

Once all the permissions were obtained, the staff began to conduct focus group discussions with different groups based on gender, age, different needs, and occupation in the community. There is a range of experiences and opinions among members of the community, and focus group discussion can help understand different perspectives on a selected topic. Once the separate sessions for the different interest groups were done then the representatives of different groups were brought together to develop community mapping and action plan.

The focus group discussions were done both in the communities and schools. All facilitators had received a training prior to their work in the schools and communities because facilitating group discussions require conflict resolution and consensus building skills besides active listening and collating the data (OxFam, 2004).

3.4.2.3   COMMUNITY MAPPING

Once the staff built rapport with the community members through key informant interviews and focus group discussions, developed community and school maps. The psychosocial community committee surveyed the community and drew a map of the area. The map included both infrastructure (communal facilities, homes, rivers, and roads) and society (community leaders, trained personnel, and people with functional needs elements, children and elderly). These maps showed the perspective of the drawer and revealed much about local knowledge of resources, land use and settlement patterns, and household characteristics. Once the maps were done, community members identified what elements of the map represented protective factors and what elements represented risk factors (Fig. 3.2).

By learning from this example, a community should note if some facilities are made before others, if some are larger in proportion than others. This will give the community members some insight into what issues may be more important than others. Take photos of the map for future references.

FIGURE 3.2   A group of community members developing a map of their community.

3.4.3   COLLABORATIVE PLANNING

When the community maps are completed and the protective and risk factors are identified, the community members are encouraged to develop action plans that will reduce the risk factors and strengthen the protective factors. This process helps the survivors to manipulate the new post-disaster reality to make their own place where they feel comfortable and secure. In other words, it helps them to reestablish their sense of place. Each action plan answered four questions:

1. *What do we want?*
2. *What do we have?*
3. *How do we use what we have to get what we want?*
4. *What will happen when we do?*

Each action on the plan turned into a community project once it answered the four questions. A project is a series of activities (*investments*) that aim at solving problems within a given time frame and location. The investments include time, money, human, and material resources. Before the community members approached Red Cross, they articulated in writing what is it that they have and will contribute toward completing their project.

This activity helps to enhance the sense of ownership of the project and its success at the community level. Projects were not approved or funded by the SLRCS/ARC if they did not directly meet the psychological and social needs of the target communities. Each community had a white board where they articulated their plan for the next 3 months with time-lines and person(s) responsible for the completion of each activity.

3.4.4 IMPLEMENTATION

The different projects were implemented collaboratively with various members in the community. Special attention was given to ensure that all people felt included and received benefits from different projects. It is important to note that some projects were targeted to include marginalized groups within the community. For example, the lace making project was developed and implemented by the woman in the community, and school chest was designed by local children and was distributed in the target schools.

The community and school leaders as well as Red Cross workers maintained logs to record the expenditure, timing of the activities, and number of people who participated or benefitted from each activity. Some projects to support the immediate community but as their benefits, the support began to spill over to nearby communities as well. For example, the block making project started to help the group to rebuild their own houses. However, once the group could complete their houses, they continued to make blocks for members of other communities. They sold the blocks for cheaper rates than in the market and generated some money. This project brought different groups of people together who took turns to make blocks while maintaining a log to record who used the machine, in-kind donation from the Red Cross, and how many blocks were produced. While they made the blocks, they shared their feelings and concerns with each other, celebrated their success together and supported each other as they walked together on their recovery path.

3.4.5 MONITORING AND EVALUATION

The project incorporated monitoring and evaluation tools at organizational and community level. At the organizational level, the team developed a logical framework where each objective and its related activities had measurable indicators attached to it. The project was implemented through six Red Cross branches and each team developed monthly and quarterly reports based on the logical framework and their work plan. There were also white boards in each target community to monitor the progress of all the community projects.

In terms of evaluation, project included a baseline survey, mid-term review, followed by some modification and final evaluation. Mid-term review was conducted with all (who could attend) the trained personnel at the community level in each target district. The structure of the review was relatively simple and included four questions: (a) what went well; (b) what needs to be improved; (c) how do we improve what needs to be improved; and (d) what are the next steps? The data was collected from all the review meetings in all the branches. Recommendations collected from the review meetings were carefully reviewed and incorporated in the program, where relevant. Final evaluation was conducted by external evaluators who used both qualitative and quantitative methodologies to collect data.

3.5 SUSTAINABILITY

The project right from the beginning built the local capacity in community-based psychosocial support approaches and tools. Hundreds of teachers, community leaders, faith-based leaders, social workers, police and military personnel were trained. Project staff also contributed to institutionalizing of psychosocial support approaches in school curriculum through the Ministry of Education. In terms of the community projects, community members were mandated to articulate their contributions right in each resilience building projects. Project budget was designed in a way where Red Cross contribution to community projects decreased overtime, encouraging community member to increase their share as the time passed. This was done so that by the time the project finished, most, if not all, activities could be sustained by community owned resources. Project also built psychosocial support capacity within the SLRC society by developing theoretical framework, project management tools such as logical framework, work plans, and training material for various tiers and training the staff.

REFERENCES

Dayal, A. A Case Study of Psychosocial Support Programs in Response to the 2004 Tsunami, 2006.

Fullilove, M. T. Psychiatric Implications of Displacement: Contributions from the Psychology of Place. *Am. J. Psychiatry* **1996,** *153,* 1516–1523.

Inter-Agency Standing Committee. Guidelines on Mental Health and Psychosocial Support in Emergency Settings (MHPSS); Inter-Agency Standing Committee (IASC): Geneva, Switzerland, 2007.

OXFAM. *Participation Tools, Including Participatory Assessment: Social Inclusion Directory;* Oxfam International: Cowley, Oxford, 2004.

Prewitt Diaz, J. O.; Dayal, A. Community Participatory Planning Processes in Psychosocial Programs, 2007. http:// www. article- hangout. com.

Prewitt Diaz, J. O.; Dayal, A. Sense of Place: A Model for Community Based Psychosocial Support Programs. Aust. *J. Disaster Trauma Stud.* **2008,** *2008*(1).

Steele, F. *The Sense of Place;* CBI Publishing Co. Inc.: Boston, MA, 1981.

PART II

Development of Community-Based Psychosocial Support After the 2005 East Asia Tsunami

PSYCHOSOCIAL SUPPORT IN SOUTH AND SOUTHEAST ASIA: FROM THE GUJARAT EARTHQUAKE TO THE TSUNAMI

JOSEPH O. PREWITT DIAZ

Chief Executive Officer, Center for Psychosocial Support Solutions, Alexandria, VA 22310, USA

CONTENTS

4.1   INTRODUCTION

The purpose of this chapter is to discuss the development of community-based psychosocial support, propose a practical and theoretical overview for the field, introduce some indicators to measure psychosocial well-being, and briefly discuss the community and school components of psychosocial support. The chapter explains how field experience, over time, diverse disasters in developing countries, and cultural, linguistic, and contextual milieus have informed the psychosocial components and approaches currently used and accepted by the humanitarian community.

The chapter introduces the theoretical framework for the development of early models of psychosocial supports. It concludes that psychosocial support is not a strategy or model that fit all contexts, cultures, linguistic backgrounds, community and individual resources, capacities, and government certification requirements. We propose that there are standard geographic settings such as communities and schools, and procedures such as psychological first aid (PFA), small projects, communication strategies and capacity building that inform program development.

The definition of psychosocial support program (PSP) has been a long-discussed issue between academics and practitioners alike. Exact definitions of this term vary between and within aid organizations, disciplines, and countries. This section introduces the early theoretical basis for psychosocial support and how it had evolved to 2010, within the American Red Cross.

In the seminal book *Disasters and Mental Health: Selected Contemporary Perspectives*, Quarantelli and Dyens (1985, pp. 158–167) introduce findings from research on community response to disaster events and draws relevant inferences about the link between the community responses and the mental health of the affected people. Early on they recognized that the "one model fits all" is not valid in disaster responses. Communities on a day-to-day basis are not structured to cope with disasters even if copious plans based on previous experience exist.

Quarantelli and Dyens suggest that "a community has to be disorganized before it can develop a new structure capable of coping with the new and often overwhelming demands made on it (p. 161)." They suggest that a disaster brings about changes in community behavior leading to psychosocial distress: (1) *production, distribution, and consumption are disrupted*: markets and stores are closed. Goods and materials are distributed at no cost, pieces of equipment are used without permission or authority; (2) *socialization is affected*: For example, school's role changes from providing formal education to children into a shelter, feeding site for survivors and workers, and a distribution center; (3) *community social control*: Laws are set aside, elected official's role changes to provide reassurance and interpretation to the affected people, task that have little to do with their job description or assigned responsibilities; (4) *social participation*: Volunteer associations assume disaster relevant activities, however, most social clubs and social association suspend their normal operation, and many cultural events are cancelled, and; (5) *mutual support*. Tremendous increase in interaction, "gallows humor" arises, and the appearance of "we" (inclusive) vocabulary is applied for those that experienced the emergency (p. 164). Because of this immediate change in everyday life, certain priorities emerge that if not handled appropriately may cause psychological stress. The learning from this study conducted more than 60-years-ago teaches us that all matters of high priority, if not dealt with well, can have both short- and long-term negative psychological consequences (p. 166).

A disaster leads to a disruption in our social perceptions of what we believe we are as a community. To recover and achieve psychosocial well-being, at least three steps have been identified: (1) a sudden change in geography, social institutions, and community culture leading to a sudden crisis in the context of change; (2) an internal tension and crisis that triggers the need to evolve, grow, and adjust. Erickson (1968) is among the first scholars to recognize and acknowledge the impact of culture, context, and human development.

Erickson's contribution to the psychosocial support work is found in his understanding and vision of human growth from "the point of view of conflicts (inner and outer), that the person manages, reemerging from a disaster with an increased sense of inner unity, good judgment, and a new vision of culture, context and community, thus, increasing the psychosocial well-being of individual and community (pp. 91–92)".

One of the tools used to understand the interaction of internal and external need for psychosocial support was developed by Erickson 1968 in his works on identity and crisis. He suggests that the concept of crisis "in a developmental connote a turning point, a crucial period of increased vulnerability, and heightened potential (p. 96)."

Bronfenbrenner (1979, 1986) conducted a systemic exploration of human development from the view of ecological psychology and determines that all individuals live at the center of a social, cultural, and ecological system. Therefore, they are impacted and they impact the social, cultural, and ecological system. Disaster-affected people experience a composite of personality characteristics (such as temperament, emotional development, coping, past experiences with traumatic events, and behavioral patterns). The model incorporates the individual biopsychological components and their interaction with other individuals, family, neighborhood, community groups, systems, and government agencies. All of these, according to Bronfenbrenner and Morris, 2006), comprise the social and ecological system that will be impacted over time by a disaster. Fullilove (1996) studied displacements in large northern cities in the United States.

She proposes that displaced people overtime are prone to develop traumatic stress resulting from the loss of physical and figurative space. She suggests that when people with symptoms expressed in the sample of her study are nostalgia.

The "sense of place" is related to the existence of dependency factors between the survivor and the environment, and signifies a dynamic process by which the survivors manipulate the environment to reflect their identities. The psychosocial interventions that attempt to reestablish the sense of place in the disaster survivors are based on the dynamic relationship between the psychological effects *(emotions, behaviors, and memory)* and the social effects *(altered relations as a result of death, separation, family, and community breakdown)*. They assist the survivors to identify their "sense of place" on the postulation, that meaning and positive emotions help to restore an individual's worldview.

"Sense of place" denotes a dynamic process by which a survivor manipulates the environment to reflect his or her identity, thus, giving an emotional meaning to the person-environment relationship post-disaster (Lazarus, 1966; Lazarus and Folkman, 1984). Loss of a sense of place leads to psychosocial distress and confusion, and the reestablishment of a sense of place leads to psychosocial competence.

The sudden onset of the disaster, such as tsunami, taxes the capacity of the community to make choices regarding their psychological and social well-being. When everything that the community networked with is gone because of the disaster, the traditional values, norms, and behaviors fail the survivor and a sense of powerlessness and "loss of place" are reported by many community members; this adds to the feeling of loss and disorientation. Survivors lose their capacity to use the resources available to them in the surroundings.

Individual psychosocial competence refers to levels by which survivors are able to guide and take charge of their lives, moving beyond the trauma. It consists of a number of factors including a sense of control and a sense of being an active part of family and community networks, engaging in active planning, and being able to manage the physical and psychological support and threat. These factors are psychosocial and interrelated in the sense that they are influenced by social and cultural factors as well as individual experiences. Each individual becomes a product as well as a contributor to the culture and its relationships to other cultures (Tyler, 2001).

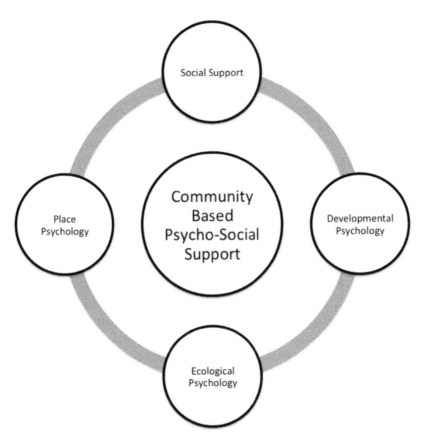

FIGURE 4.1   American Red Cross early model of community-based of psychosocial support 1998–2004 (Prewitt Diaz, 2006).

A review of existing disaster mental health and psychosocial care programs outside of India indicated that existing programs throughout the world today are predicated on the stages of the continuum of disaster. A lot of emphasis is placed on the development of community-based programs that use community resources and encourage the integration of community networks.

4.2   THE DEVELOPMENT OF PSYCHOSOCIAL SUPPORT IN SOUTH INDIA

4.2.1   PSYCHOSOCIAL SUPPORT IN SOUTH ASIA

Psychological support program in India and South Asia were initiated in 2001 as a result of the Gujarat earthquake in Bhuj and the communal riots in Ahmedabad. The American Red Cross International Services assigned Dr. Gordon Dodge to conduct an initial assessment in conjunction with a colleague from the International Federation of the Red Cross and the Red Crescent (IFRC). They identified the geographical area impacted and the immediate psychological needs.

Dr. Gerald Jacobs took over during the recovery. He hired a local doctor as the local counterpart and developed a plan very similar to the domestic interventions on disaster mental health in the United States with the concurrence of the IFRC and the Indian Red Cross Society (IRCS), as well as WHO. The plan was submitted to the American Red Cross Headquarters, where funding was held up until a long-term delegate that Ms. Marcia Kovacks took over as the psychosocial delegate and moved the agenda forward by socializing the disaster mental health concept with the IRCS.

In February 2002, the Gujarat riots in Ahmedabad caused the death of over 2000 Muslim, abuses of women and children, and massive destruction. The delegate was moved to New Delhi for security reasons. In May 2002, the IRCS requested a delegate to train a group of volunteers to provide assistance to the affected population in Ahmedabad. Dr. Joseph O. Prewitt Diaz, who had gained experience in India during a short assignment during the Odisha Super Cyclone in 1999, was sent to Ahmedabad for three weeks where he trained a group of 12 volunteers, and developed, gathered data through narratives, and focused groups in the centers served by the IRCS. He prepared a proposal to assist this population, and it was funded partially by ARC-India.

In the meantime, Dr. Srivasa Murthy and Dr. K. Sekar from National Institute of Mental Health & Neuro Sciences (NIMHANS) had developed, since the Bhopal deadly gas leak, a psychosocial care approach that addressed secondary needs of survivors with long-term interventions. Their teams provided assistance to survivors of the Gujarat earthquake and the communal riots in Ahmedabad. An association was formed between NIMHANS, Care-India, the American Embassy, and the IRCS with community groups. The project was called Project Harmony and the objective was to bring all segments of the community back together.

After the departure of the Ms. Kavacks, the psychosocial delegate, Dr. Prewitt Diaz, was assigned to this post in Central America to India in 2002. His tasks were to develop a program in India based on the assessment findings made by his predecessors, visit affected areas, and the intentions of the IRCS to institutionalize this program. It was a seemingly daunting task to set up a psychological support program in a country as culturally diverse and geographically spread out as India. This challenge became an opportunity to plant and sow the seeds for one of the most successful International programs of the American Red Cross. After that two major initiatives took place in Gujarat and Odisha.

4.2.2 THE INDIAN RED CROSS SOCIETY/DISASTER MENTAL HEALTH AND PSYCHOSOCIAL CARE PROGRAM

Following the Gujarat earthquake in 2001, the IRCS requested the American Red Cross to provide technical assistance in the development of a community-based program that will provide immediate assistance to survivors of the disaster and first responders, and provide assistance after a disaster in the form of disaster preparedness in schools and villages in selected states. Unless otherwise indicated the chapter relies on the book entitled *Disaster Mental Health in India* (Prewitt Diaz et al., 2001).

A review of existing literature in India (Suri, 2000; Mukherjess, 2002; Kumar et al., 2000; Bharat et al., 2000; Sen Dave et al., 2002; and Sekar et al., 2002) presents an in-depth discussion on the need for psychosocial care after disasters such as the Odisha super cyclone, the Gujarat riots, and the Bhopal gas leaks. It is the agreement of these studies compiled by Laksminarayana (2003) that during the immediate response there are great needs in the population and that many government and nongovernment organizations provide support to the surviving population.

There is concurrence that psychological support is a long-term proposition that should be addressed by government organizations in the case of individual needs and nongovernment organizations in terms of community-level work. The IRCS, on request of the Government of India, provided assistance in the form of meeting the basic needs of survivors who were living in camps following the Gujarat earthquake. The results suggest that the immediate assistance was helpful to the survivors, and recommend that long-term assistance be a pursuit to achieve psychosocial well-being in the affected population.

A planning meeting was held on October 28–29, 2002 by the IRCS with representatives from target states and partner organizations with the objective of developing guidelines for a community-based psychosocial support program. A (strengths, weaknesses, opportunities, and threats) SWOT analysis looked at strengths, weaknesses, challenges, and opportunities of the IRCS as it related to the existing disaster mental health program.

The greatest strengths of the IRCS were its representation in all states and its volunteer force of eleven million members, which placed it at a huge advantage in delivering community-based programs. The greatest weaknesses were the (1) lack of trained personnel in Disaster Mental Health or psychosocial support at the national, state, and local branch level, (2) lack of appropriate materials for preparedness activities in schools and villages, and (3) lack of integration of PFA with the existing community-based first aid programs already in existence within the IRCS.

During the meeting, we learned that the Ministry of Health was responsible for providing health services to the population including mental health services. A challenge for the Ministry which became evident during the Gujarat earthquake was that (1) there were no real health posts to serve all the villages affected by the disaster and that (2) those villages that have health posts they were staffed by personnel who would conduct primary health activities and then refer to a hospital. (3) A National Health Response Plan during a disaster does not include community-based mental health.

The experience of the partner organizations such as the National Institute for Mental Health (NIMHANS), Oxfam India, ActionAid India, and CARE in developing community-based mental health programs in affected communities in Odisha and Gujarat indicated that psychosocial care is an important part of the long-term rehabilitation of the survivors. These organizations reported that (1) proactive community interventions are important, (2) community-level helpers are an important link to provide service, (3) care should be provided to the total community, (4) practical assistance is as important as emotional support, and (5) psychosocial care should be a long-term proposition.

The planning meeting suggested that there is currently no integrated plan in India to provide disaster mental health or psychosocial support to survivors of disasters. The opportunity for the IRCS was to develop a community-based disaster PSP that complements the services offered by the Ministry of Health and that it is a span of services following the guidance in the SPHERE project.

4.2.2.1 PRINCIPLES OF THE INDIAN RED CROSS SOCIETY COMMUNITY-BASED PSYCHOSOCIAL SUPPORT STRATEGY

The IRCS looked at the best practices in psychosocial support after a disaster or complex emergencies. The IFRC has developed guidelines for psychological support (Abdallah, and Burnham 2001; IFRC Working Group 2001). These guidelines have been elaborated in consultation with the International Working Group for psychological support, which met in Geneva, 5–8 April 2001. The working group identified principles to be valid for preparedness as well as during the disaster.

The policy for psychological support (2003) stated that (1) it is the responsibility of the national societies to include psychological (in 2004 the focus changed to community-based psychosocial support pursuant the guidance in SPHERE project 2004) perspective in every area of intervention, (2) design psychological support as a component in other programs like disaster preparedness, disaster response, first aid, health, social welfare, youth and organizational development, (3) provide psychological support as a long-term and reliable commitment to ensure that the psychological aspects of relief work are professionally implemented and make a crucial difference to the population, volunteers, and staff affected by the disaster. The IFRC (2003) policy refers to the design of psychological support according to the basic principles. The basic principles defined by the IFRC and adopted by the IRCS are discussed below.

4.2.2.1.1 A Community-Based Approach

As opposed to a clinical and/or individual approach; the majority of reactions following a disaster, for example, distress and suffering are not psychiatric illnesses (and *do not*, therefore, require professional treatment), but are reactions that can be prevented from developing into something more severe if services such as information, psychological education, and support groups are provided. Working with people on an individual basis should be the exception, as this only responds to the needs of a few, and might lead to stigmatization. Tackling problems in isolation is expensive and is not sustainable.

Any organized activity should relate to everyday realities and priorities that have been identified by the communities. It is important to make use of institutionalized social infrastructure already in existence. Target beneficiaries of PSP should be considered for active survivors rather than passive victims.

4.2.2.1.2 Community Volunteer Technicians and Specialists

With training and support from mental health professionals, community volunteers can work in an independent, efficient, and effective manner. These community volunteers have the access to, and the confidence of, the beneficiaries. And, importantly they benefit from the necessary cultural sensitivity to provide adequate assistance to the affected population.

4.2.2.1.3 The Program will be Technically Appropriate and Sensitive to the Cultural and Linguistic Diversity of India

Programs should be designed and implemented through a continuous community dialogue. The goals are: (1) to reintegrate individuals and families within the community and identify and restore natural networks and coping mechanisms, (2) be aware of the degree of heterogeneity of the community, (3) communicate with local partners and beneficiaries to create mutual respect for linguistic diversity and cultural beliefs and the expertise of the staff and local professional resources.

4.2.2.1.4 Identifying and Strengthening Problem-Solving Resources in the Community

The IRCS recognizes the importance of the communities' capacity to help themselves through their own support networks and coping mechanisms that existed prior to the disaster. One of the important tasks pre-disaster is to find out about the communities' previous and existing coping mechanisms and strategies, and support or build on the same. In some situations, support structures may have disintegrated as a consequence of the disaster, and an alternative structure has to be introduced. This new structure should be adapted to the community's pre-disaster traditions. Facilitate access to communication with family and relatives and to a family reunion, because these are very effective methods in promoting psychological well-being, and in reassuring people, especially children. Focus on people's positive efforts to deal with and come to the terms with their experiences, but without minimizing their concerns.

Based on the principles outlined above and best practices proposed in the literature, the experience of our partner organizations in India and the identified community needs, the Indian Red Cross Society (IRCS) planned and began to implement a community-based PSP. The goal of this program is to *"Alleviate stress and psychological suffering resulting from disasters."* The program broadly focused on four key areas namely: (1) capacity building of national headquarter (NHQ), state and local branches; (2) expand the capacity of trained community volunteers to offer PFA; (3) develop "Safe Schools" through creative and expressive activities and crisis response planning, and (4) recognize and enhance the resilience of communities through resilience building activities.

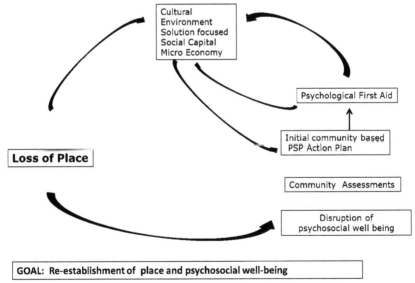

FIGURE 4.2   Procedure of reestablishing place utilizing community-based psychosocial support. (Figure developed by Prewitt Diaz for this article)

When a community is impacted by a disaster or an emergency, the loss is not only emotional but also has other components. There is a loss of natural and built environment, an immediate discrepancy emerges between what were the cultural norms, and how they have been modified (e.g., the relations and roles change within the family, the importance of the elders, traditions, and customs in handling these type of events). To assure safety, security, and comfort there is a need to engage in solution-focused exercises (i.e., what do we have, what do we need, how do we get it), the need to identify through community assessments, what material, intellectual and physical capital exist in the community, and what types of assistance is needed externally to breach gaps. What people-intensive, small projects can be developed to engage community members in their own recovery, assuring that through community-based psychosocial support, the community reestablishes place and assures psychosocial well-being.

By the beginning of 2003, the psychosocial program was developed in two target states Gujarat and Odisha based on the existing theoretical and practical views expressed above. Three hundred selected schools and villages will develop disaster response plans with a focus on stress mitigation during and after disasters. It was designed to complement the services provided by the Ministry of Health and the partner organizations during and after disasters at the state level and national level.

4.2.3   CAPACITY BUILDING OF STATE AND NATIONAL HEADQUARTER

The biggest challenge, in the beginning, was to improve the capacity of the target states in developing culturally appropriate systems, methods, and materials that reflect the reality of India. The American Red Cross experience in the domestic disaster services and in the South and Central America was used as the foundation to set up the structure for a viable program.

At the community level, the community facilitators were similar to a community volunteer, the Crisis Intervention Technician developed knowledge of PFA, community engagement and mapping, and principles of recovery. The crisis intervention specialists (CIS) were either a psychologist or social workers with 3 years of experience. The overall program coordinator at the NHQ level would be called the Crisis Intervention

Professional and have at least a Master's in Public Health, an MD, or Ph.D., and 5 years of fieldwork. The overall capacity building objective of the program was to form a Crisis Response Team at the state and NHQ.

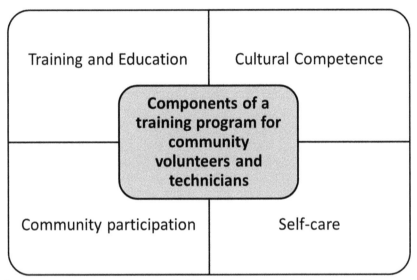

| Training and Education | Cultural Competence |

Components of a training program for community volunteers and technicians

| Community participation | Self-care |

Figure 4.3 Components of capacity building and psychosocial technicians. Volunteers are spontaneous personnel recruited in the affected communities. All psychosocial personnel will bring linguistic, cultural, and contextual competence. They will participate in capacity-building activities that will create or enhance skills in PFA, community mobilization, and assessment, exhibit inclusive behaviors. They will receive in-depth training in sense of place theory and identify, instruct, and practice strategies of the reestablishment of place. All personnel will be literate in techniques of self-care and have the capacity to provide peer-to-peer counseling.

The American Red Cross provided technical and financial support to the IRCS so that the IRCS would build the capacity of the national society by conducting staff development activities in all sectors: NHQ, state branches, and local branches. Direct intervention benefitted the target schools and villages. The skills of teachers and community facilitators were enhanced through the capacity building using thematic units focusing on group development, community-based psychosocial support preparation, promotional activities, and skills necessary to actively promote and encourage proactive behaviors.

The capacity-building effort was multipronged focusing on psychosocial workers and professionals at various levels. The compulsory follow-up for any training activity is at least 100 hours of supervised practice in the field. Capacity building was defined in a tiered system beginning with basic knowledge and practical skills about the Red Cross, psychosocial support, PFA, and community mobilization activities. Capacity building blocks are introduced below.

4.2.3.1 OPERATIONAL TRAINING

The operational training for Red Cross volunteers, community volunteers, and teachers is an 18–32 hour workshop to help participants to implement short-order interventions (such as PFA) and activities in the communities and schools. It is a brief introduction, which provides the requisite knowledge and skills to carry out PFA, focused groups, community assessments, and creative and expressive activities with children in schools and communities (See Fig. 4.4 below). These workshops were followed by practical experience of up to 100 hours.

Aimed at reducing disaster related emotional distress

- Develop skills in Psychological first aid
- Identify indigenous psychosocial support practices
- Establish mechanisms od support with the Health Posts
- Assist in identifying social capital
- Assist in Identifying and mobilizing community-based re-establishment of place strategies.
- Assist communities in develop Strategies to adapt to the new place

Capacity building for volunteers: Psychosocial Building Blocks

FIGURE 4.4   The figure introduces the objectives of thee capacity-building activities for the two initial tiers of training: (1) community facilitators and (2) technicians.

4.2.3.2   COMMUNITY FACILITATORS

The Psychosocial Support Training for community facilitators aims at developing and enhancing the skills of community leaders, volunteers, and other community members to carry out community-based psychosocial support activities. The training followed by field supervision develops skills such as PFA, participatory community assessments to ascertain the impact of the disaster on natural and built environment, access of marginalized persons, and social capital. In addition, they advise and assist the community to conduct cultural events and promotional activities, as well as assist in organizing solution-focused community resilience activities. Community facilitators are the links between the community and external stakeholders including the Red Cross. They liaise with all groups in the community to include the elderly, women, adolescents, men, diverse cast, disabled, and elderly. A total of 120 community facilitators were trained in Bhuj, Gujarat and 147 in Odisha between 2003 and 2004.

4.2.3.3   TEACHERS TRAINING

The psychosocial support training for teachers was designed to prepare teachers to carry out creative and expressive activities within the curriculum, to facilitate children to express their feelings after a disaster, crisis, or emergency and to organize students, to facilitate volunteers, parents of students, and school personnel to prepare a school crisis response plan. Schools are frequently the most stable institutions in the communities, and teachers are among the most influential and respected persons in the community. Teachers are well prepared, to direct the preparation of three dimensions of community maps, and facilitate problem-solving sessions. With children, they provide the conduit for conducting place-attachment activities, through writing and drawing, and reestablishing cultural and environmental place.

The next two tiers of capacity building were composed of crisis intervention technicians and specialist. The core responsibilities of these volunteers and paid staff were to: (1) provide supervision and support to the activities of community volunteers; (2) assure the validity of methods used in conducting community assessment; (3) coordinate with external stakeholders', including the Red Cross, plans and proposals to reestablish place; (4) mobilize social capital and make sure that the resources are used appropriately, and assist in planning for, and reconstructing natural and constructed environment, as a small community project or cash for work activity.

4.2.3.4   CRISIS INTERVENTION TECHNICIANS

The crisis intervention technicians have responsibility for a team of up to five volunteers (although this responsibility may increase with the spontaneous increase of volunteers during a given period). The technicians were full-time volunteers or paid staff at the local branches of the IRCS. This group of individuals is trained at the state level; therefore, they learn to work closely with peers from other Red Cross branches and to develop state-wide psychosocial support teams that come together in the event of a major state-wide disaster. They are responsible for developing training volunteers and teachers in PFA, community mobilization to reduce distress, and psychosocial place reestablishment activities in communities, villages, and schools. Most technicians identified themselves as local branch IRCS volunteers, school teachers, or others in the community with experience of disaster response.

4.2.3.5   CRISIS INTERVENTION SPECIALISTS

The CISs serve as managers' planners and supervisors of a group of up to five technicians. The specialists are selected among the teachers, psychologist, social workers, or community health workers with over 5 years of experience. The capacity development consists of a 15 days' residential and a 300 hours supervised fieldwork certification training program. The capacity-building activities were designed to enhance skills of the trainees to supervise, coach, plan, and design and implement psychosocial support interventions in schools and communities. Each trained CIS is required to serve 300 volunteer hours in a period of 1 year to receive certification. These voluntary hours include providing supervision, psychosocial support, and training technicians and volunteers.

In the preparedness phase, the specialists were involved to provide education about psychosocial support to the schools and community. These CIS personnel were members of the crisis response team and coordinated planning and response to immediate crisis intervention and PFA activities to the survivors and emotional support activities for the first responders.

The CISs were prepared at the state level. Odisha has a total of 62 trained CISs providing 300 hours of volunteer service to the Odisha State Branch and Social Welfare Advisory Board. In Gujarat, there were two courses held and a total of 30 CIS who provided 300 volunteer hours to the Gujarat State Branch.

The capstone capacity building activity was called the Crisis Intervention Professional course. This was a by-invitation course that lasted 30 days and required 500 of certified experienced persons in psychosocial support activities during a disaster. There were 12 CIPs trained serving in the IRCS (three of these were employed by the America Red Cross-India delegation).

4.2.3.6   CRISIS INTERVENTION PROFESSIONALS

The crisis intervention professionals are individuals who have a background in mental health or disaster preparedness, response, or management. They were assigned to the National Society and committed 500 hours of volunteer time after completing the program of study. The prerequisite to participate in this highly-specialized program was a Bachelor or Master's degree or higher and/or equivalent experience in disaster-related activities.

The role of the professionals was to advise the psychosocial support teams of the IRCS on program planning and implementation before, during, and after a disaster, and conduct or coordinate the development and execution of a rapid needs assessment in the affected geographical area, call out other mental health professionals to augment the response of the Crisis Response Teams, organize psychosocial response teams and coordinate teams in the field in order to provide timely service to the affected geographical area and also develop and conduct capacity-building activities. The professional will also proactively collaborate and network with other Governmental and Non-Governmental Organizations involved in psychosocial to influence and establish policies related to psychosocial.

At the onset of the 2004–2005 tsunami response, it was this team of professionals hired as third-country delegates by the American Red Cross, that orchestrate the psychosocial response and recovery activities between 2005–2008 in the Republic of Maldives, Sri Lanka, and Indonesia. The original planner was an Indian psychosocial support professional (PSP), with the support of a health delegate that developed the original psychosocial response plan adopted by the International Federation of the Red Cross (IFRC). The two immediate responders, a psychiatrist and a social worker (PSP's), originally assigned to the IRCS National Headquarters in New Delhi, and to the state branch in Odisha.

4.2.4 MATERIAL DEVELOPMENT CENTER

One of the challenges for psychosocial support response is the assumption that there is going to be one set of guidance notes that will apply universally if they are translated to the target language. For example, in shelter management, we know that a person will have a 15-ft square to sleep, or in the water and sanitation guidance, a person should receive 15 L water a year (SPHERE project, 2011). The funding agency allows few resources material development and beneficiary cost per person for distribution of preexisting psychoeducation materials for translation and copying to the target language.

There is a limited guidance of the transference of knowledge and information from one language, culture, and context to another. The need for a material development center emerged as we attempted to respond to transferability of knowledge, and concept development to the Indian context, multiplicity of language and existing knowledge. Initially, we thought that translation of the words into the local languages would suffice. When pamphlets and handout included pictures, we found out that the pictures would impact the message as received by the affected population. The other communication challenge was to develop messages that were easily understood by the target population. Thus, the emergence of a material development center happened in India to serve populations in Hindi, Gujarati, Uria, and Katchi. The center prepared materials that were culturally, linguistically, and contextually appropriate. All materials were developed focusing on the concept of transfer of knowledge, changing attitudes, and modifying behaviors (KAB). Over a period of 6 years, Indian material development center published hand-outs, tri-folds, cultural and contextually appropriate nonverbal tools, short videos, and books.

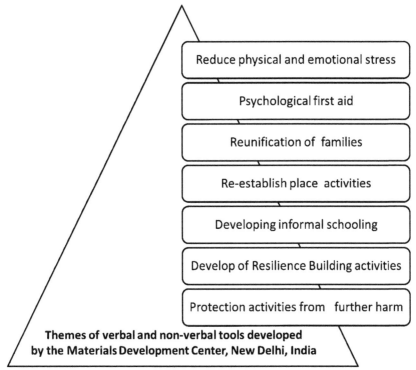

Reduce physical and emotional stress

Psychological first aid

Reunification of families

Re-establish place activities

Developing informal schooling

Develop of Resilience Building activities

Protection activities from further harm

**Themes of verbal and non-verbal tools developed
by the Materials Development Center, New Delhi, India**

FIGURE 4.5 The figure presents the seven themes used in developing verbal and nonverbal materials in New Delhi, India. With the tsunami response, this technology was transferred to the Republic of Maldives, Sri Lanka, and Indonesia.

The material development effort of the IRCS PSP was to prepare the target villages and schools to recognize potential psychological risks among members (anxiety, fear, and traumatic stress), and the ways to address them so that they would be reduced. The manuals helped teachers and community facilitators to identify what positive strategies had to be developed, and practiced, to cope effectively with the loss and reestablishment of place. The materials included (1) "share your feelings" trifolds and posters; (2) PFA trifolds, posters, and manuals; (3) community-based psychosocial support manuals and nonverbal tools for community facilitators, (4) school-based psychological support manuals for teachers (elementary, intermediate, and high school), (5) psychosocial intervention sessions for children manual and nonverbal tools, and sets of share your feelings cards to assist in community mapping as (6) an assortment of tools.

The instructional materials were technically, linguistically, and culturally adapted from several sources including already existing American Red Cross materials. The teachers' training curriculum for the Safe School program was adopted from the "Master of Disaster" curriculum and Central America Mitigation Initiative material on a community-based psychosocial support, the community facilitators' training modules were adapted from SPHERE project, The United States Agency for International Development (USAID), UNICEF, and the PSP center in Copenhagen.

These manuals, trifolds, handouts, nonverbal tools, and activity books prepared by the program were field-tested, translated, and contextualized into six Indian national languages. Based on the lessons learned and field experiences a second edition of the manuals was developed. Some of these materials were included in the national curriculum (CBSE) of India in 8–10th standard textbooks. The trifolds and posters were used during the Bam earthquake in Iran wherein WHO recognized ARC staff for contribution in the development of the Iran National Psychosocial response.

In the aftermath of the tsunami, many of these materials and, certainly, the cultural and linguistic equivalence were adopted by the used in the preparation of materials in Maldives, Sri Lanka, Indonesia, and India. Some of the drawings in the nonverbal tools have found their way into the international scene through the development of the IASC mental health and psychosocial guidance.

During the tsunami response, these materials were adapted to the local settings and translated into Dhivehi and Singhalese in the areas of operation, that is, the Maldives and Sri Lanka. This has been the biggest achievement of the program wherein a standardized set of PSP materials were prepared that were technically appropriate for any part of the region and were easily adapted into the area of operation.

4.2.5 SAFE SCHOOL PROGRAM

The safe school program is aimed at enhancing the skills of teachers, students, and volunteers to prepare for and respond to disasters, crises, and emergencies through training, crisis response planning, creative and expressive activities. The training includes both psycho educational materials and resilience-building activities that help children to express themselves within the classroom. The Safe School program impacted 899 teachers and 31,382 children in the India program

4.2.5.1   *SCHOOL CRISIS RESPONSE PLANNING*

To enhance resilience and build the capacity of schools to respond to disasters or emergencies, it is important for teachers, students, parents, and volunteers to be involved in psychosocial preparedness activities. The crisis response planning is a process that facilitates the participation of the school community in taking proactive measures to develop actions/ steps in the event of a disaster. It involves the formation of four teams namely, evacuation, damage assessment, PFA, and physical first aid, and organizing simulations every 3 months so that the school can practice the planned steps. In Gujarat, 49 schools developed crisis-response plans and conducted PFA and first aid trainings for the school children and teachers.

4.2.5.2   *CREATIVE AND EXPRESSIVE ACTIVITIES FOR CHILDREN*

This is the crucial aspect of the Safe School program which involves engaging teachers in a process wherein they can understand the causes of stress in children and the most common reactions that are seen in class as well as at home. The use of creative and expressive activities to facilitate children to express their feelings after a crisis or a difficult experience helps to create a nonthreatening environment and encourages children to talk. The program includes enhancing teachers' skills in using some of the creative and expressive techniques, such as drawing, clay, drama, and play, within the classroom. School chests and recreational kits consisting of teaching aids and play materials are distributed to participating teachers following a short operational training on how to use creative and expressive activities within the classroom.

4.2.6   **SAFE HOMES PROGRAM**

The program is designed to prepare community-level volunteers designated as "community facilitators" (in the ratio of 1:100 population) from communities that are vulnerable to disasters. Community facilitators are provided with some skills that help facilitate community in psychosocial activities, promotional activities, and mobilize the community to develop community crisis response plans and develop psychosocial support brigades.

The primary focus of the safe homes program is to recognize and enhance the psychosocial well-being of the community. This involves constant dialogue and focuses group discussions with the community members to assess and understand the inherent coping mechanisms, traditional methods of healing, and community cohesion.

4.2.7   FORMATION OF CRISIS RESPONSE TEAMS

The intent of the CRT is to supplement local resources with guidance and leadership from a national or state team of specially trained and experienced Red Cross responders. The concept also includes a commitment to mentoring others and building the capacity of local branches to respond to future incidents. The capacity building within the PSP over the past year led to the formation of crisis response teams in Gujarat and Odisha, capable of providing immediate support to survivors of the disaster. These teams were mobilized by the Indian Red Cross during the school-lightening incident in Chaupalli.

The PSP adapted the suggested activities set forth by SPHERE Project (2004), such as, engaging adolescents and adults in reorganizing their community with cash-for-work activities, engaging traditional healers and respected community leaders in the community-grieving activities, and using school in Odisha and the Kumbhakonam school fire in Tamil Nadu (see next section of the paper for further description of these interventions).

4.2.8   SUMMARY

The IRCS PSP intervention has become a classic response that is utilized as a case study by the National Disaster Management Institute. This intervention provided three lessons: (1) moving IRCS PSP volunteers from other states for 3 weeks to assist the survivors with the community and school programs was challenging but worthwhile; (2) capacity building can effectively take place with the assistance of a wide variety of community stakeholders including Red Cross volunteers, teachers, and crisis intervention technicians, and (3) the IRCS' "peer to peer" provided a good example of how psychosocial support helped its own human resources to begin to go forward in the process of psychosocial support and assist the affected people to move from victims to victors.

The DMHPC program was designed to generate skilled volunteers at different levels. Networking with Government and Non-Government agencies was a crucial component of the program in order to support and sustain the pool of skilled mental health workers developed by the program. Agencies such as The United Nations Development Programme (UNDP), the Odisha State Disaster Mitigation Authority, the Ministry of Home Affairs of the Government of India became some of the agencies that the program liaised with during the 2 years.

At the end of 2 years, the program reached out to over 2 million direct recipients not only in the target states of Gujarat and Odisha but also in Andhra Pradesh and Tamil Nadu. Training at the different levels strengthened volunteer force of the Indian Red Cross with a commitment toward over 17,000 hours of service.

4.3   FROM THEORY TO PRACTICE IN ACTUAL MAJOR CATASTROPHIC EVENTS

4.3.1   HIGHLIGHT OF POST-EARTHQUAKE IN GUJARAT, INDIA

Coastal villages in four districts in the state of Odisha: The super cyclone in 1999 had caused extensive destruction in the tribal villages of coastal Odisha. Thousands of people had lost their family and friends and all of their life's savings. Rehabilitation and reconstruction were still ongoing 3 years after the disaster. Partner organizations such as Action Aid India and NIMHANS were carrying out psychosocial programs in one of the worst affected districts.

A participatory assessment was conducted in 30 villages along the coast, and it was found that emotional distress and startled reactions were still present in the community. The community expressed the need for psychological support, thus, the program was initiated in 30 villages initially and was expanded to other districts. Schools and community programs were supported around the Chilika Lake. By 2004, the communities had addressed their personal distress and communal differences that resulted due to issues related to the super cyclone. They were to establish a market to sell their agricultural products and were able to initiate tourist trips around the lake. The communities used the shelters that had been constructed by the German Red Cross. They held meetings on the lower bottom floor and used some of the shelter space on the second floor as a community school; in the morning for small children and in the afternoon for teenagers, learning trades.

4.3.1.1   BHUJ EARTHQUAKE SHELTER (GIDC), BHUJ, GUJARAT

A population of 10,000 people continued to live in temporary housing, 12×12 ft shacks made of tin, 2 years after the earthquake. The survivors were living in small colonies divided along caste and religious lines. The living conditions were extremely unhygienic due to the lack of basic services such as safe drinking water, drainage, and electricity. Although, the people living in such conditions were still reliving the past, undergoing stresses of daily living, as well as planning for their future in terms of resettlement and livelihoods. Local mental health and psychosocial personnel were well known in the community but had been denied access to the refugee camp by the local authorities. In 2003, the Red Cross visited the community and was not well received, after two meetings between the volunteers and the community members.

The breakthrough happened, and the community committee expressed the urgent need to provide preschool for children in their respective sectors. The community was willing to provide a tin shed if the Red Cross would assist with salary for a teacher, an aide, and a cook. One week later 14 such preschools were functioning in the GIDC. A participatory assessment with shelter leaders and members from different community groups revealed the need for psychological support primarily for children and elderly. The program was started with active support from the community. Components of a training program for community volunteers and technicians.

The program continued for 5 years and it promoted cultural activities, training, capacity building, and supported three schools in the camp (Government school special needs children, the Muslim School, and a Catholic school that provided K-3 schooling). In January 2008, many families had reestablished their place, however, many of the special needs people remained in the GIDC, and to that date, they were still using techniques and activities to support their psychosocial well-being. An important achievement of the Gujarat program was the inclusive organizing interventions and activities that reached out to all the caste and religious groups in the shelter and marginalized groups.

The following community psychosocial activities were conducted in the GIDC shelter in Bhuj during 2003–2004.

- The facsimile of a day-care center for the children aged 2.5–6 years old was supported (10 centers). A team of community facilitators supported the program. Nonformal education methods were used in the centers to teach the children and a mid-day meal was provided. The program was successful in reunifying and integrating orphan children into family-like environments.
- The program facilitated the community-based integration of about 200 adolescents from different castes, nonscheduled caste, the Muslims, and the Christians by holding focused groups pertaining to adolescent issues, organizing sports activities and other campaigns that make the youth feel being a useful part of the community. Computer classes were supported to help adolescents to enhance their skills and develop resilience by enabling them to pursue alternative means of livelihood.
- Women in the shelter were seen facing two great problems—they had lost their dear ones and did not have necessary skills to earn their livelihood. Some 497 women in 10 block groups were selected as the target group. The program emphasized on skills development such as sewing, silver-ware, tie and dye work, Mehendi (art form performed in a person's hand and feet).
- Community volunteers supported the elderly people to organize programs that will make them self-sufficient (arts and crafts, painting, day trips, and monthly routine check-ups). Approximately 211 elderly persons benefited from these activities.

4.3.1.2   ANJAR SCHOOL IN ANJAR, GUJARAT

The Disaster Management Authority declared that the city had suffered 100% damage during the earthquake. The initial psychosocial assessment showed that 2 years after the earthquake, the remnants of the destruction of the natural and built environment was a reminder in the minds of the survivors of the losses they had incurred. Children and parents were still coming to the terms with the loss of family and loved ones. One of the schools had lost 189 children and 22 teachers on that day as they were parading to the nearby ground to celebrate the Republic Day. Based on the dialogues with teachers, community members, and survivors, 10 schools, and 10 communities were selected in Anjar for psychosocial support interventions. In 2005, all the 179 teachers had developed their capacity to provide PFA to the students. They had developed specific didactic materials in the Katchi language to share their feelings. The communities had conducted their assessment and requested funding to provide assistance to the elderly and to the people who had experienced physical damage caused by the earthquake. In 2010, the project was still going strong with the support of the personnel trained in the local branch.

4.3.1.3   CRISIS RESPONSE TEAMS

By the 2nd year, in 2004, the IRCS requested the development of PSP crisis response teams. Two brief case studies emerged from some of the experiences. Both disasters occurred in school but left a deep emotional scar in the affected communities.

Chaupalli school lightening on 22nd August 2004, Odisha, India: A primary school in Chaupalli, Odisha (a rural area near the tribal area) was struck by a lightning, that left one child dead, injured 37 children and one teacher and the entire school in shock. Immediate assessment by the PSP rapid response team of the Odisha State Branch, augmented by two National Staff, showed that children had feelings of fear and powerless/helplessness, and were experiencing traumatic stress as a result of the event. Many of the children were refusing to attend, and most parents preferred their children staying at home. The entire community was in fear of stepping out of their homes when it rained or sending their children to school.

A team of ten people, led by two ARC staff was deployed in Chaupalli for a period of 10 days. The team met the community, listened to the adults, and explored the village with villagers about what had been previous experiences with violent weather. The community members provided some examples of how they coped in the past, and what were some activities that would help them cope in the future. By the end of the intervention, the community had developed a plan that was written and shared with every household. Psychoeducation in pictorial form was shared with all the residents.

Creative and expressive activities were conducted with the school children and teachers, and community volunteers were oriented on using the same activities in the course of their classroom activities. Each classroom was given a school chest and a recreational kit. The children were given breakfast, drawing, explaining their drawing, creative physical movement, and manual arts. At the end of the day, the children would receive lunch and were dismissed. This intervention lasted 5 days. At the end of the intervention, the attendance fluctuated from 63 on the initial day to 78 in the last 3 days. The teachers continued an abbreviated version of the activity for 6 weeks by utilizing community volunteers to carry on the activities.

Kumbhakonam school fire, 16th June 2004. The Kumbakonam school fire tragedy of July 2003 was one of the worst human-caused disasters of the decade in Tamil Nadu. As a result of the fire, 93 children died in an elementary school in Kumbakonam, which is in the Tanjabur district of Tamil Nadu. The cause of the fire was accidental, but the children's deaths were a result of teachers' unplanned actions and a lack of disaster preparedness.

In addition, 18 children were severely injured and 784 school children in the community were psychologically affected by the incident according to the initial assessment by community volunteers. The Joint Secretary of the Indian Red Cross sought technical assistance from the American Red Cross and mobilized the National Psychosocial Response Team to support the affected children and their families.

In the immediate aftermath of the fire, the basic response was to identify the bodies, proceed with funeral rites, and then begin to deal with the issues of loss. For many families, they suffered not only the loss of a child or children but also material losses from expenses incurred in medical treatment and burials. The initial psychosocial response by the Red Cross personnel was to focus on meeting the basic needs of the survivors which, in this case, was shelter, food, and physical health.

A team of two people from the IRCS Bhuj program accompanied by a Crisis Intervention Professional from the ARC India delegation were mobilized to conduct an in-depth assessment and respond accordingly. An operational training of a group of local community volunteers in PFA and rapid needs assessment was conducted. These volunteers were native speaker of the local language (Tamil) and knowledgeable of the local culture and context. Most of them had a Bachelor's degree. The needs assessments revealed that five communities and schools reported psychosocial support needs. Referrals to the local hospital were made of persons that were experiencing stress reactions, reported difficulty in handling the loss of their children. The community volunteers began working with families in five communities in providing PFA and organizing community-based interventions to facilitate the expression of feelings and conducting creative and expressive activities with children.

By the end of November, a group of 21 local volunteers was trained in assessment, PFA, and community and school-based psychological interventions. The Kumbakonam subbranch of the IRCS was developed and a group of 30 educators trained to continue to support the children's psychosocial needs. In addition, the communities had organized themselves and began to develop a disaster response plan. The response lasted for 12 months.

The responses offered by the community during the focus groups' assessment activities informed the development of the long-term psychosocial program. While the American Red Cross (CIP) lead remained in Kumbakonam for a period 5 months, the IRCS CISs rotated every 3 weeks to provide maximum exposure to a disaster response to these personnel. The IRCS specialist, with technical assistance from the American Red Cross CIP, provided the following technical assistance during the 5-month intervention period: (1) sensitized all segments of the community about the need for psychosocial support for the survivors of the Sri Krishna school fire accident; (2) built the capacity of 18 volunteers with capacity-building activities and assisted the communities of Kurrupur, Asoor, Prumbandi, and Old Pallakarai in organizing themselves for the period of 12 months intervention, and conducted three capacity-building activities on the "Safe School program," and each classroom received a school chest with sufficient materials for 1 year. The schools that got the benefits from Safe School program were (1) Saraswati Primary Pathshala, (2) Banadurai Primary School, and (3) Nattham School.

4.3.2   TSUNAMI IN INDIA, MALDIVES, AND SRI LANKA, 26TH DECEMBER 2004

The rapid response teams developed and trained with technical and financial assistance from the American Red Cross were equipped to handle response activities in psychological support. This team of personnel served as response teams in Tamil Nadu, Andhra Pradesh, Maldives, and Sri Lanka during the tsunami operation of both the Indian Red Cross and the American Red Cross. They were able to use the experience and training developed during the India program and culturally adapt the same according to the country.

The tsunami occurred during the Winter break for the India Delegation of the American Red Cross (December 24, 2004). One of the most powerful earthquakes on the planet hit the Asian Continent with its epicenter at Aceh, Indonesia. Some of the tsunamis reached as far as 1600 km (91,000 miles) from the epicenter of the 9.0 magnitude quake, which was located about 160 km (100 miles) off the coast of Indonesia's Sumatra Island at a depth of about 10 km (6.2 miles).

There was only one delegate, a CIP, who lived in New Delhi on the ground when the tsunami occurred. The day after the tsunami the Country Head of Delegation instructed the CIP to develop a psychosocial support response plan and get it ready to present the IFRC field assessment and coordination team (FACT) already on the ground in the Maldives. A health delegate had flown into New Delhi to work in tandem with the CIP to develop the plan. The IFRC had shared this initial report:

Initial Report on Psychosocial Needs in the Maldives
From the mental health point of view, the country has one psychiatrist and one clinical psychologist for the 300,000 population and they are on the capital island of Male. Since transportation is a big challenge in the Maldives, practically mental health services in the islands are nonexistent. The awareness of mental health need was rudimentary and people did not recognize it as a need till the tsunami. The concept of community and preventive psychiatry did not exist. Currently, the existing system for psychological support is overwhelmed and community-based psychosocial support programs with rapid resource generating capacity are the need of the hour (IFRC/FACT Initial Report. December 26, 2004. Male, Maldives)

On December 27, 2004, the FACT of the IFRC identified psychological needs of the affected population and contacted the American Red Cross Asia regional delegation based in India for sending teams for immediate response to the psychological needs of the community. Based on the call and needs identified by the FACT psychosocial support team was dispatched from India to implement immediate psychosocial response programs in the Maldives.

The following needs were identified by the FACT:

1. *Loss:* The tsunami took many lives in the Maldives, destroyed or damaged many homes and infrastructure, and seriously affected economic livelihoods. Other home losses included furnishing and other belongings, for example, clothing, personal items, identity cards, birth and other official certificates. Because most of the islands do not have any banking system, many affected families also lost whatever savings they had kept in their homes, and now find themselves completely without identity, autonomy, and what psychologists term "loss of place." The implications of all of these are likely to be far-reaching for physical health, psychosocial health, and well-being and steps need to be taken to address the situation quickly.

2. *Direct psychological morbidity:* Direct psychological morbidity having various degrees of severity appears to be prevalent in the most acutely affected islands and in less impacted but close by neighboring islands. A recent pilot survey suggests that prevalence may be high, but the extent to which trauma will persist and prove functionally disabling is not clear. It is also noteworthy that chronic anxiety appears to have been a major problem among communities even before the tsunami and may have been indicative of poorly managed transitioning from highly traditional and relatively isolated society and culture to a tourist-dominated, the one that exposed people to other ways of life and behavior as well as to a better income. Based on the findings of the local psychosocial counseling teams supported by the United Nations Fund for Population Activities UNFPA as well as discussions with both displaced and non-displaced people, people who felt they had lost everything and others who had lost relatively little, it seems as if anxiety, fear, hyperarousal, feelings of insecurity, sleep and eating disorders, irritability and loss of future orientation are currently common. While none of these reactions is necessarily abnormal over the short term (and in the case of heightened alertness could be positive), prolonged sustained fear, nightmares, intrusive thoughts and hyperarousal could become highly dysfunctional at both a clinical

and public health level. This is of special concern given the poor psychological support infrastructure outside of the main atoll, Male.

3. *Indirect psychosocial problems:* The presentation of psychosocial problems varies. Many of them are to be expected and are probably transitory. Other symptoms, however, are being noted and there is a growing incidence of more chronic and serious problems including symptoms of traumatic stress and other problems. These could go on to have long-term psychological and social impact, and be associated with socially dysfunctional behavior as well as with personal and familial problems. There is also a good reason to believe that as time passes and the unpredictability of the future begins to become a concern for many of the internally displaced people (IDPs), the risk of further psychosocial difficulties and disability will become pronounced. What form these difficulties and disabilities will take is not clear, but it could easily assume a form of more generalized functional disorders. If this occurs, the capacity of society as a whole to rebound and reconstruct itself will be threatened. Because these kind of problems have never been evident in a large proportion of the population before, there has not been a major investment in developing the type of system and infrastructure needed to deal with such problems.

4. *Displaced person camps:* Many displaced persons feel increasingly isolated and neglected. In some cases, they feel that they are purposely not being involved in activities and information sharing. Some seem to be withdrawing and beginning to respond passively to attempts to help them. Other psychosocial issues involve displacement of adolescents (and indeed younger children) who have lost their education records and are now not clear what will happen to them. Coping is taking a variety of forms and public displays of sociosexual behaviors that were previously not typical are occurring on some islands. Other potentially problematic manifestations, which are readily evident in some of the displaced populations, are a lack of activity, listlessness, and serious environmental untidiness. The Government is taking steps to manage the situation and is setting up shelters on 6 atolls, but the national authorities lack some of the technical expertise and knowledge required for management of the shelters from a psychosocial perspective to nurture an environment of the reestablishment of place. Capacity

needs to be strengthened for the management of IDP settlements and ensuring the health and protection of women and adolescents if not; human security and especially psychosocial well-being will be threatened and the chances of sustainable social reconstruction will be constrained.

The CIP in New Delhi developed an initial response plan and dispatched a two-person team (a psychiatrist and a social worker) with the task of; (1) training a group of 60 counselors in community-based approaches such as PFA; (2) continuing the psychosocial assessment; (3) developing local capacity, (4) have culturally and linguistically appropriate intervention.

Three sessions of PFA training had been provided by the end of January 2005 to 59 counselors, which included the counselors under the SSCS (mostly), doctors and nurses from the government hospital (which was catering to the injured from the islands), and police personnel (who were accompanying the counselors to the islands and were first responders).

The PFA module was a 2-day module developed and field tested extensively in India (1200 PFA trained in different parts of India). The module was translated and linguistically, culturally, and contextually adapted by a group of three local mental health professionals. Feedback was obtained from the three training sessions. Only minor modifications were required due to the cultural similarity between Maldives and India. The final version of the manual is still currently used by the Maldives Red Crescent Society.

The second team with two CIPs was dispatched from New Delhi to assist teachers to develop a Safe School program in all the atolls. The module for Safe Schools was translated in culturally adapted with the help of local teachers. With a capacity of six teachers from the Maldives, a team was built to carry out teachers training in the local language under the supervision of a trainer from American Red Cross. A short training for immediate intervention was given to teachers in all islands of the Maldives for intervention in the children who account for more than 50% of the population. The Education Ministry expressed the initial need for an operational training to enable teachers to handle the immediate reactions of children in the island schools before the school reopening. A collaborative and participatory process was adapted to define a method of fulfilling the need.

A planning meeting between the representative of the Education Ministry and the Psychological Support Delegate and training team was held to develop an appropriate implementation strategy keeping in mind the logistics and travel considerations in the islands. Based on lessons learned from the psychological support program in India and discussions the strategies were (1) build the capacity of a group of six teachers from schools in Male; (2) develop a 2-day operational training for teachers based on the American Red Cross "managing a disaster" curriculum and the 7 day school based psychological support training manual, (3) prepare three teams consisting of two teachers and one member from the training team as technical assistance to conduct the trainings in the target atolls; (4) the target atolls were selected on the basis of the number of schools that were scheduled to reopen. The manual was titled "Operational Teachers Training Manual for Tsunami Affected Area." It consists of four modules and annexes containing lesson plans for resilience-building activities.

The six volunteer teachers were committed by the Education Ministry to travel to the islands for the duration of the teachers' trainings. These teachers were able to carry out the second phase of the trainings independently in four teams. The fourth team consisted of two teachers who received the teachers training.

The goal of the sessions was to establish that children have the capacity or resilience to recover from this and other events and the role of the teacher in building that capacity. The participants' own capacities were explored and shared. They felt that the key factor that makes Maldivians resilient was the fact that all belonged to the same religion and spoke the same language. They identified that their society was the relatively conflict-free one that contributes to a feeling of oneness and motivates them to help one another in times of crises.

Lessons learned in this first international response to the crisis intervention professional, recruited as third country delegates by the American Red Cross is presented below:

1. *The utilization of local teachers* instead of trainers from the ARC training team as core trainers to deliver the course facilitated two processes. First, it augmented participation of the island teachers. This was because the training was delivered by their peers from Male. Second, it led to the linguistic and cultural adaptation of the training course and manual.
2. *The practicum on day 2* where the trainee teachers conducted some of the activities learned in the session with children of the affected

islands served two purposes. First, the participants had the opportunity to practice the theory. This accelerated the learning and clarity because they could see the effectiveness of the activities when used with children. Second, they had the opportunity to provide direct intervention for the children from the affected islands. This helped to motivate them and increased their belief in the concepts taught in the training.

3. *Involving teachers from the island schools by the trainee teachers* during the visit to affected islands for the practicum contributed to building a network between the teachers. It facilitated the sharing of the knowledge received and inculcated a commitment to pass on the information to peers and colleagues in their own schools.

4. *The used method of capacity building* proved extremely efficient in preparing a core group of teachers who are now able to deliver the course independently without the support of the ARC training team. This method was focused on a small group of six people and the skills were built through modeling and constructive feedback by the ARC trainers.

4.3.3   INDONESIA PSP PROGRAM

The earthquake that measured 8.9 on the Richter scale struck the area off the western coast of northern Sumatra on December 26, 2005, triggering massive tidal waves or tsunami's that swept into coastal villages and towns in the Aceh provinces as well as the surrounding countries. As a result, more than 210,000 persons lost their lives, causing major damage to the coastal province of Banda Aceh. The population of Aceh is approximately 4 million. Banda Aceh, the capital city, has a population of 230,000. The population along the severely affected western coast of Aceh was approximately 450,000. Virtually all of those living in Banda Aceh and the western coast died or were displaced.

The hardest hit areas were the provinces of Aceh and Northern Sumatra in Indonesia. In many areas families have been left homeless, children have been separated from their parents, family members are missing or dead and there is an urgent need to address the emotional, physical, and social needs. In addition to the immediate toll on human lives, the long-term effects on infrastructure, facilities, the economic well-being of communities, will continue to plague the Indian Ocean rim countries for years to come.

At the heart of the quake, Aceh and Northern Sumatra provinces in Indonesia are devastated. Out of the total population of 15.5 million of which 5.5 million are children (estimates based on the most recently available official government figures 1999) are living in these provinces. The culture and community, had strong resources such as family networks, community networks, and religious practices that are mobilized for comforting and coping with the situation. The uprootedness, devastated homes, and loss of life caused by the disaster, many families and individuals are affected by the disaster emotionally, physically, and socially.

PSP was conducted in the areas selected by the National Society. It provided assistance and supported the affected people of the Tsunami disaster who lost family members, friends, social, and economic livelihood to increase their capacity and alleviate immediate psychological effects, reducing the risk of long-lasting mental disorder.

Psychosocial activities were designed to assist the affected people to come to terms with their loss and build upon the capacity to withstand adversity. The program identifies the factors in the survivor's cultural and social environment that support the capacity to withstand and cope with adversity. The program will reinforce the individual survivor and address the social relationships in schools and communities.

The American Red Cross deployed a PSP Delegate for a 4-week mission between February and March. The objective was to support members of the Palang Meera Indonesia (PMI) (Indonesia Red Cross) to develop the capacity of volunteer and paid staff in Banda Aceh working with families during and after mass grave burials. The second intervention began in June of 1968 with the deployment of two CIPs from New Delhi to Banda Aceh, as third country delegates to the American Red Cross to develop a PSP in Banda Aceh, Aceh Gaya, and Calang. Palang Merah Indonesia with assistance from the PNS psychosocial support consortium including ICRC (International Committee of the Red Cross), Turkish Red Crescent Society (TRSC), Australian Red Cross (AuRC), American Red Cross (AmRC), Japanese Red Cross (JRC), Hong Kong Red Cross (HKRC), and Danish Red Cross (DRC) as the lead agent of the consortium.

External coordination meetings were conducted with UNICEF and Save the Children. In May 2006, the Turkish Red Crescent completed their intervention and left Banda Aceh. The American Red Cross took over two camps with displaced widows and their children, ten communities and the Madrazas (school and Community Center on the grounds of the mosques).

The CIPs identified volunteers from the Banda Aceh Branch of the Indonesia Red Cross that would have an interest in participating in the program. Two fluent speakers of English and two psychologists were hired in the Banda Aceh Red Cross Office. An immediate assessment of the affected communities assigned to the American Red Cross was conducted in target villages and schools.

The first volunteer team is stationed in Banda Aceh conducted community assessments, networking with other NGOs, supporting local volunteers and explored local health referral systems. The local Radio station is used to raise public awareness of stress and sports activities arranged for the local volunteers.

By the end of August 2005, the psychosocial staff had begun with the following activities:

1. Capacity building of the PMI regarding PSS to develop the quantity of trained personnel, and scope of psychosocial support in 30 branches, Rapid Response Teams PMI and the PMI National Headquarters in Jakarta. These programs target mental health professionals, Teachers, schools, and nonprofessionals. The different capacity-building programs have a range from 18 to 36-hour training program with 200–500 hour of obligatory voluntary work included attached. (1) Crisis intervention specialist, (2) community-based disaster mental health/psychosocial care training for community facilitators, and (3) school-based psychological support training for teachers.

2. By the end of the 2nd year of the program (August 2007), there were 997 identified community and school volunteers. Support for capacity building of volunteer became a step process from community facilitator to Crisis Intervention Professional. Six seminars for PSP training and mobilization of volunteers took place in Medan with 168 volunteers. The program was implemented in Aceh Jaya—Teunom, Calang, Krueng Sabee, and Lamno. The target groups were children, adolescents, adults, elderly, and local

volunteers were mobilized from the communities where the activities are being implemented.

3. By the end of the 2nd year of the program (2006), there were thirty (30) volunteers enrolled in a Masters' degree program with the University of Indonesia in Jakarta. This cadre completed their program of study in December 2009. Most of these personnel are part of the Branch staff and still work and the reestablished communities. Their presence in community clinics and school has helped in continued development and improvement of psychosocial support activities to keep schools and communities safe.

4. Developed community-based approaches for psychosocial support, with the active participation of the local community that were culturally, linguistically, and contextually appropriate following the "Safe Community" model developed in India. Assisted in the reconstruction of "Medrazas" (community centers) to facilitate community linking and a community social meeting space.

5. Selected and developed formal and informal education activities in conjunction with UNICEF. These activities included reconstructing and reequipping target schools. Conducting informal schooling in the madrasas (schools located in the grounds of the mosque). In the morning, these centers were occupied with young children as early childhood centers. The Red Cross had 10 psychologists working within 16 camps in the Logan district in Banda Aceh, who are trained in PSP and are targeting schools, villages, and camps. Psychosocial support was delivered to target areas by the 10 psychologists. These personnel made referrals to the mental health hospital.

6. Psychologists disseminated information through trifolds and small group meetings on topics related to the loss of family, friends and the natural and built environment. Psychoeducation through radio and TV spots about the importance of engaging in daily routines. At the end of the 2nd year of the program, 3000 teenagers and children had been involved in recreational and sports activities in their communities, games, and drawings. The PSP conducted activities in 12 Islamic orphanages working with boys and girls children and teachers that were not affected by the disaster.

7. Developed facsimiles of the "Safe Schools" program to build the capacity of teachers, children, and their parents in the basic of PFA and psychosocial support. More than 200 such programs were

developed successfully. They were different in their interest, and day-to-day activities, however they were well organized to take care of it other in the case of a stressful life event.

8. Crisis Response Teams were organized within each branch of the Red Cross in Aceh, Aceh Gaya, and Calang. These teams proved their value during the Yogyakarta volcano eruption and earthquake. Three local CRT teams deployed to the affected areas worked in tandem with the medical teams, and alleviated fear among the affected people for a month after the event, helping many to return to normalcy in the schools and communities. These were self-supporting, trained, and capable to move at a moment's notice.

4.3.4   INDIA PROGRAMS AFTER THE TSUNAMI

Two programs were initiated after the 2004 tsunami, planning and execution was conducted by the American Red Cross CIP's that had returned to form the regional response, this cadre of professionals had developed sophisticated skills to move affected people from the traumatic stress caused from loss of life, property, and community "loss of place" to a community-based approach where psychosocial support became the platform for community recovery and reestablishment of place. The clinical approaches adapted in the early 21st century interventions, regenerated into a community-based approach, where psychosocial support became the platform for a large group of affected people to get together figure out what they lost through mapping exercises and using solution-focused strategies identify what they need to reestablish themselves. The evaluation of both programs in Kanyakumari, Tamil Nadu, India, and Anjar, Gujarat, India, concluded with the following observations:

1. The affected people began to identify reestablishment of place activities. They began to feel ownership over the process of recovery and participated in various planned activities.
2. The active involvements of the affected people in community participatory process made them aware about their own capacity to survive their losses of place, and solution-focused behaviors that helped to reestablish themselves individually and as a community

are based on the available resources. Furthermore, it generated trust amongst the affected people.

3. The community centers were dynamic, functional, and active with the interest of the community members. It was about where the people would congregate—a crossroad, a local bus stand, a place of interest for most affected people, or a school building or community center.

4. Community members became knowledgeable about the simple interventions of PFA and began to practice the basics of self-care activities (rest, eat, talk, socialize, and work together).

5. The intervention through nonformal school increased the opportunities for the children and adolescents to become more secure about their personal skills and learn new problem-solving skills.

6. The community planned interventions increased the sense of psychosocial well-being among the villagers.

To paraphrase a metaphor from Fullilove (2016), "root shock occurred, at first sight, all was lost, the survivors were uprooted. Like the fallen tree with its exposed roots, that withered away as though dead, one day sometime later, a little sprout shows, then a leaf, and a young tree begins to grow, from the roots of the fallen tree, it grows bigger and stronger over time." Quarantelli and Dynes (1985) in early research found that those who stay near their home, near their "place" are more prone to recover and rebuild than those that have been uprooted to another "unfamiliar" place. Psychosocial support as a platform works on the psychological hole left by the uprooted tree, rather than the unfamiliar relocations of an unfamiliar location.

Psychosocial support as a platform supports generating an integrated perception of needs, identification of social capital, and technically sound interventions managed by the affected community, and accountable to all stakeholders. The strategies are informed focused groups that are held in functional community centers and are transparent and rely on community human capital. Where needed, there may be agreements with external stakeholders. As the community reestablishes there a place there is evidence of increased trust and cohesion among the affected people, stronger human capital, greater community knowledge, practices, and skills.

4.4 SUMMARY

This chapter focused on the experiences of developing psychosocial support strategies over time, in several South Asian countries and different disaster settings that were contextually, culturally, and linguistically diverse. The program acknowledged the need to promote reestablishment of place.

By December 2006 most of the American Red Cross CIPs, having completed their 2-year tasks, returned to India. Having built the capacity of paid staff and volunteers transitioned unfinished components of the PSP activities to the skilled hands of national societies. The learnings were significant for future program development: (1) to be successful on the ground and psychosocial support programs must be community driven: identified, planned, executed, and celebrated; (2) constant institutions in the community reduce the fear of the community and increase the desire to reestablish place; (3) for external collaboration it is essential to coordinate efforts with external and internal stakeholders; and the assumption that there is a timeline for recovery and is erroneous, it may take years for a community to address loss and decide to reestablish itself such was the case with the community of people that suffered physical trauma, and (4) capacity building from the neighborhood volunteers to the psychosocial professional is paramount. New skills and practice sharpen the volunteers' skill to better assist the affected people to care for themselves.

REFERENCES

Aarts, P. G. H. *Guidelines for Programmes: Psychosocial and Mental Health Care Assistance in Post-disaster and Conflict Areas;* International Centre: Netherlands Institute for Care and Welfare: Utrecht, The Netherlands, 2001.

Barton, R. Psychosocial Rehabilitation Services in Community Support Systems: A Review of Outcomes and Policy Recommendations. *Psychiatr. Serv.* **1999,** *50,* 525–534.

Bradburn, N. M. *The Structure of Psychological Well-Being;* Aldine Publishers & Co.: Chicago, Illinois, 1969.

Bronfenbrenner, U. *The Ecology of Human Development: Experiments by Nature and Design;* Harvard University Press: Cambridge, MA, 1979.

Bronfenbrenner, U. Ecology of the Family as a Context for Human Development: Research Perspectives. *Dev. Psychol.* **1986,** *22*(6), 723–742.

Bronfenbrenner, U.; Morris, P. A. The Bioecological Model of Human Development. In *Handbook of Child Psychology;* Dammon, W., Lenner, R. M. Eds.; John Wiley & Sons: New York, 2006; 6. Vol 1, pp 793–828.

Coelho, G.; Hamburg, D. A.; Adams, J. E.; Eds. *Coping and Adaptation;* Basic Books: New York, 1974.

Crocq, L.; Crocq, M. A.; Giapello, A.; Damiani, C. In Lopez-Ibor J. J., Christodoulou, G., Maj, M., Sartorious, N., Okasha, A. Disasters and Mental Health, New York: John Wiley and Sons, 2005.

Erickson, E. *Identity, Youth and Crisis.* Norton Publishers: New York, 1968.

Fullilove, M. T. Psychiatric Implications of Displacement: Contributions from the Psychology of Place. *Am. J. Psychiatry* **1996,** *153*(12), 1516–1523.

Fullilove, M. T. *Root Shock: What Wee Can Do About It;* New Village Press: New York, 2016.

IASC/MHPSS. *Guidelines for Mental Health and Psychosocial Support in Emergencies;* Interagency Standing Committee: Geneva, 2007.

INEE. *Minimum Standards for Education in Emergencies;* Interagency Network for Education in Emergencies: Paris, France, 2004.

Jacobs, G. The Development of a National Plan for Disaster Mental Health. *Professional Psychosocial Res. Pract.* **1995,** *26*(6), 543–549.

Jahoda, M. *Current Concepts of Positive Mental Health;* Basic Books: New York, NY, 1958.

Martel, C. *Quebec's Psychosocial Interventions in an Emergency Measure Situation;* Ministry of Health and Social Services: Quebec, Canada, 2005.

Mascelli, A. American Red Cross Disaster Services. In *Mental health Response to Mass Emergencies;* Lystad, M., Ed.; Brunner/Mazel Publishers: New York, 1988; pp 181–210.

McFarlane, A. C. Stress and Disaster. In *Extreme Stress and Communities: Impact and Intervention;* Hobfoll, S. E., Vries, M. W., de Eds.; Kluwer Academic Publishers: London, 1999.

Morgan, J. Providing Disaster Mental Health Services through the American Red Cross. *NCP Clinical Quarterly* Spring, **1994,** *4*(2), 3–6.

NSW Institute of Psychiatry. *Disaster Mental Health Response Handbook;* British Columbia NSW. State Health Publications: Paramatta, 2000.

Psychosocial Working Group. *Considerations in Planning Psychosocial Programs;* Institute for International Health and Development: Edinburgh, United Kingdom, 2004.

Quarantelli, E. L. An Assessment of Conflicting Views on Mental Health: The Consequences of Traumatic Events. In *Trauma and Its Wake;* Figley, C. R. Ed.; Brunner/ Mazel: New York, 1985.

Quarantelli, E. Ten Criteria for Evaluating the Management of Community Disasters. *Disasters,* **1997,** *21,* 39–56.

Quarantelli, E. L.; Dynes, R. A. Community Responses to Disasters. In *Disasters and Mental Health: Selected Contemporary Perspectives;* Sowder, B. Ed.; U.S. Department of Health and Human Services: Washington, D.C., 1985.

Raphael, B. *Disaster Management;* Australian Government Publishing Services: Canberra: Australia, 1993.

SPHERE Project. *Humanitarian Charter and Minimum Standards in Disaster Response;* The Sphere Project: Geneva, 2004.

Tyler, F. B. *Cultures, Communities, Competence and Change;* Plenum Publishers: New York, NY, 2001.

WHO. *WHO Assessment Instrument for Mental Health Systems;* WHO: Geneva, 2005.

Young, B.; Ford, J. D.; Ruzek, J. I.; Friedman, M. J.; Gusman, F. D. *Disaster Mental Health Services: A Guidebook for Clinicians and Administrators;* Department of Veterans Affairs: Menlo Park, California, 1998.

CHAPTER 5

DEVELOPMENT OF A LINGUISTICALLY APPROPRIATE INSTRUMENT IN TWO STATES OF INDIA

SATYABRATA DASH[1], ANJANA DAYAL DE PREWITT[2], AND RASHMI LAKSHINARAYANA[3]

[1]*Senior Psychiatrist, Apollo Hospitals, Bhubaneshwar, Odisha, India*

[2]*Associate Director Psychosocial Support Program, Save the Children USA, Washington, D.C., USA*

[3]*Senior Consultant, EY K.S Institute of Technology, Bengaluru, Karnataka, India*

CONTENTS

5.1   INTRODUCTION

The purpose of this chapter is to report on the development of a measure of resilience for two States in India. A functional definition of resilience is the ability to bounce back from an adverse situation and successfully rebuild yourself and your place.. It is not a characteristic that happens by chance, it emerges in people who have worked hard to regain social and psychological competence and who have the cognitive skills and strengths to overcome challenges. Psychological resilience (Rutter, 1987) is the person's capacity to avoid a psychopathologic state despite difficult circumstances and to withstand stressors without manifesting psychological dysfunction such as persistent negative moods or mental illness.

Tedeshi et al. (1998) suggest that knowledge of an individual's and community's resilience and vulnerability before a disaster, as well as understanding their social and psychological responses to such events, enables the psychosocial personnel to promote resilient and healthy behaviors that sustain the social fabric of the community and facilitate recovery. The resilience theory explains how some people are more likely to engage in health-promoting activities in the schools and communities and reestablish their lives and their surroundings. Much of the research examines the interaction among factors that protect and nurture, along with conditions in the family, school, and community, in young people's lives, which help them adapt positively after experiencing a disaster.

Understanding the community as a psychosocial entity permits the long-term rehabilitation and sustainability program planners to assist the community in identifying the factors that will enhance resilience and promote well-being. It helps the people to overcome the loss and grief and resume normal activities.

There are interacting psychological and social factors that assist people to overcome adversity. Psychological factors include self-esteem and self-confidence, internal locus of control, and a sense of life purpose. Social factors include social support from family, friends, teachers, and community.

A caring family that sets clear rules and standards, strong bonds with an attachment to the school community, and relationships with peers help people recover and reestablish a new sense of place (Luthar, 2000; Kass, 1998; Blum and Reinhard, 1997; Luthar and Ziegler, 1991; Rutter, 1987). Bernard (1991) suggests that social competence, problem-solving skills, autonomy, and a sense of purpose are the characteristics that are consistently exhibited by resilient young people.

In the final analysis, emotional maturity is the one factor that is most likely to characterize resilience in the face of disaster (Valliant, 2003). Situations of major loss demand the ability to give up previous certainties and consider the survival of others more important than the discomforts and suffering of one's immediate situation, which is essential to embody a sense of hope.

There has been an increase in post-disaster studies that attempt to measure why some people recover faster than others. The studies have identified three constructs that explain the capacity to bounce back, these are: (1) a person has the support of parents, friends, and other adults; (2) a person has spiritual and psychological resources that will help her or him move ahead; and (3) a person has the desire to move ahead and improve his or her condition.

The theoretical base for the initial activity was the resiliency paradigm. The three paradigm components are the individual's feelings that: (1) I have the support of someone in my immediate environment, (2) I am responsible and respectful of myself, and (3) I can figure out ways to solve problems (Artz et al., 2001). Begin sentence with Encouraging and cultivating encouraging and cultivating community resilience is a positive or healthy approach to reduce risk factors prior to a crisis, an emergency, or a disaster.

Psychological support as a response to a disaster or a traumatic life event has been plagued by diverse approaches The purpose of this study is to determine a valid and reliable way to measure resiliency in communities affected by natural disasters in India. An increasing body of research from the fields of psychology, psychiatry, and sociology gives evidence that most people can recover from risks, stress, crises, and trauma and experience life success (Resiliency in Action, 2002).

The construct of resiliency has its origins in the self-efficacy theory proposed by Bandura (1989) described a person's internal belief that one could accomplish a specific task. Some people believe that they can master a task because the have the capacity and desire. This belief leads to success even in challenging circumstances. These people can see themselves succeed in spite of great challenges or great adversity, such as the aftermath of a disaster. Bandura theorized that setbacks and difficulties in person's lives teach them that success usually requires sustained effort. The factors identified by Bandura are best identified as protective factors. Protective factors are individual or environmental safeguards that enhance a person's ability to resist risks and foster adaptation and competence.

Resiliency can be described by individual, family, and community level, each interdependent and complementary of the other levels. This usage implies a track record of successful adaptation in the individual who has been exposed to biological risk factors or stressful life events, and it also implies an expectation of continued low susceptibility to future stressors (Werner and Smith, 1992, p. 4). Family resilience is composed of characteristics, dimensions, and properties that help families to be resistant to disruption when major changes or crises occur (McCubbin and McCubbin, 1988, p. 247).

Within the context of this study, self-efficacy and psychological hardiness are the factors referred to as resiliency. Grotberg (1995) has conducted an international study to determine the elements of resiliency. She defines resilience as a "universal capacity which allows a person, group, or community to prevent, minimize, or overcome the damaging effects of adversity."

Emotional resilience comes from developing (1) curiosity about the future; (2) flexibility, resiliency, and adaptability; (3) a pattern of learning from unpleasant experiences and seeing victims become victors; (4) strong self-esteem; and (5) an expectancy of good outcomes. Bandura (1989) suggested that once people realized that they had the capacity to succeed within them, they proceed to tackle even the most difficult situations and to persevere in the face of great adversity and rebound back after great adversity. By persevering, people are able to emerge from difficult situations with a stronger sense of efficacy. Ultimately, the most positive thing psychosocial staff can do after a disaster is to assist people to realize their psychological hardiness and to help them define the degree of control and commitment that they still possess.

After a 2-year period of gathering data in 14 countries from 589 children and their families or caretakers, Grotberg's research grouped the main factors that influence the behaviors of survivors into three categories. "I have," "I can," and "I am." The instrument reported herein used the three categories suggested by Grotberg to develop a 25-item instrument that will assist the program staff to identify resiliency makers in selected villages and communities in two States of India.

5.2 METHOD

5.2.1 STUDY 1

One hundred and twenty people participated in this survey in Bhuj, Kutch, and Gujarat. Villagers from two villages, a Hindu village and a Muslim village, were selected. The participants were divided into 48 females and

72 males with an age range of 15–45 years. All the responders had experienced the earthquake. Most of the responders could not read or write.

The instrument consisted of three open-ended questions:

1. Who is the most supportive person in your family and in the community for you? What makes this person important to you?
2. What are the characteristics you have that helped you to survive such a major disaster?
3. When something difficult happens to you, can you express yourself to others and find ways to solve the difficulties that arise?

The recorders had a written questionnaire from which they asked the questions to subjects. The recorders read the questions to the responder and made a verbatim notation of the answers. All the interviewers brought back the responses and handed them over to the experimenter.

A panel composed of three persons (one psychologist and two teachers) evaluated the answers to every question. The judges were briefed about the meaning of resiliency as presented by Bandura (1989) in terms of self-efficacy and as defined by Grotberg (1995). The judges generated a total of 155 potential stems in 6 different categories related to resilience: (1) ability to be happy and contented and have sense of direction and purpose; (2) capacity for productive work, and a sense of competence and environmental mastery; (3) emotional security and self-acceptance, self-knowledgement, and a realistic and undisturbed perception of oneself, others, and one's surrounding; (4) interpersonal adequacy and a capacity for warmth and caring relating to others, and for intimacy and respect; (5) desire to serve others; and (6) religious, cultural, spiritual affiliations.

5.2.2   STUDY 2

Ten mental health professionals, utilizing the Brislin method, were asked to translate the 155 items obtained from the Gujarat questionnaire from English to Urya (Prewitt Diaz et al., 2003). First, the items in English were translated into Urya and then a back translation into English was done. A content analysis of the original translation and the back translation were compared to determine contextual applicability. A panel of four judges evaluated the stems and made the necessary changes to accurately represent context, meaning, and technical adequacy intended in the original items.

A second panel composed of five mental health workers were given a definition of resiliency provided by Grotberg (1995) and asked to determine whether the items represented a part of the definition of resiliency. Each stem was ranked on a 5-point scale where 1 represents not related, 2 represents poorly related, 3 represents related, 4 represents very related, and 5 represents the definition of resiliency. Seventy-six items were retained and ranked. A rater reliability analysis was performed with the rater reliability determined by a Kuder–Richardson formula 21 (KR-21), the reliability was 0.944. Items with an inter-item correlation of 0.40 and higher were retained and included in the preliminary version of the resiliency scale. A total of 52 items were retained.

5.2.3   STUDY 3

A final version of the resilience questionnaire that consisted of 52 items was administered to 900 subjects in Orissa, who were survivors of the Orissa super cyclone. The Lickert scale consisted of five choices. The scores were coded from 0 to 4 and were mentioned below each question number.

The sample for this study was selected utilizing the random cluster method. A total of 188 villages were affected by the Orissa super cyclone. The sample consisted of men and women between 18–50 years of age.

The criteria for participation in this study were that the respondents had lost an immediate relative during the super cyclone and were now experiencing success.

Thirty clusters were identified utilizing a table of random sampling. Each examiner identified the center of the cluster, spun a pencil on a map, and walked to the limit of the village and then began to zigzag through the village picking every third house until 30 subjects had been interviewed.

The rest of the participants were inhabitants of the villages served by the Orissa State Branch of the Indian Red Cross Society Cyclone Shelter Program. The instrument was administered orally by community mobilizers, who wrote down the response given by the participant.

In the initial resiliency instrument trials, when the instrument was administered in its totality and the questions were not divided into subscales, the interviewers faced difficulty in achieving survivors' cooperation. There were criticism and reluctance from the survivors because they already had been a part of methodical surveys by various government and nongovernment organizations.

In addition to the above constraint, the survivors were confused about the content of the questions and complained of repetition when different questions on the same topic were asked at intervals. It was difficult for them to appreciate the different aspects of the same area of inquiry.

Sensitivity to the survivors' emotional and cultural context was given paramount importance to achieve cooperation of the individuals assessed. The questionnaire was administered in a way that made the assessment procedure a more supportive and continuous conversation rather than a methodical survey. In addition, measures were taken to keep the survivor focused on the context of the questions asked so that they could understand the different aspects that each question intended to cover.

To achieve the survivors' cooperation, psychosocial staff divided the questionnaire of 52 questions into eight subscales that assessed different areas of functioning and support. The questions in the subscales covered the individual's dynamics of interaction at self, family, social system, and occupational environment levels. The subscales were:

1. Qualities of the individual
2. Feelings
3. Use of humor
4. Spiritual belief
5. Social support system
6. Involvement with the community
7. Capacity to work
8. Pattern of working

The answer sheets were collected and revised. Answer sheets with incomplete answers or those lacking the basic demographic information were deleted from the statistical analysis. The total number of subjects finally used for this analysis was 668.

5.3   STATISTICAL ANALYSIS

An SPSS statistical package was used to analyze the data. The statistics were determined for the 52 items. An intercorrelation analysis was performed. The items were grouped into three factors "I am," "I have," and "I can." The items with an inter-item correlation of less than 0.40 were rejected. The final instrument was composed of three subscales and a total of 25 items. Appendix A below presents the statistics for the resilience scale.

5.4   RESULTS

An instrument was developed to measure resiliency in a group of subjects in the states of Gujarat and Orissa in India. The original instrument was developed in the Gujarati and Urya languages. A translation into English has been prepared for purposes of reporting the data. The instrument was administered to a group of 668 subjects between the ages of 18–27. Initially 52 questions were developed. The stimuli questions were translated by utilizing the translation and back-translation methodology (Brislin, 1981). The final questionnaire consisted of 25 items divided into three categories (I am, I can, and I have). The alpha coefficient of reliability was (25 items) 0.94. This instrument, in its preliminary form, may be used by planners of community-based psychosocial support programs to establish baseline data and measure the degree of whether the interventions increase the level of self-reported resilience.

APPENDIX A

TABLE 5.1   Category, Mean, and Standard Deviation for items of Resilience Scale N = 668.

Category	ID	Item	Mean	Standard Deviation
I can	1	When I make plans, I can follow through with them.	1.9222	0.6409
I have	2	I have been successful all of my life.	1.5389	0.7468
I can	3	I can do many things at a time.	1.5225	0.7673
I can	4	I can succeed because I have experienced difficulties before.	1.6662	0.7352
I can	5	I can look at a situation in more than one way.	1.5674	0.7407
I can	6	I can take things one day at a time.	1.3757	0.6998
I can	7	I can find something to laugh about.	1.7171	0.7197

TABLE 5.1   *(Continued)*

Category	ID	Item	Mean	Standard Deviation
I have	8	I have people in my community who love me.	1.9461	0.6049
I have	9	I have several who listen to me.	1.9805	0.5932
I have	10	I have someone to help me when I am in trouble.	1.9865	0.6084
I am	11	I am part of at least two community groups.	2.1332	0.7297
I have	12	I have a network of people who believe that I can succeed.	1.8353	0.6838
I can	13	I can excel in creative activities.	1.6811	0.6674
I have	14	I have the capacity to achieve my goals.	1.8428	0.6599
I am	15	I am listened and respected in my community.	1.8847	0.6618
I am	16	I am valued by my friends and neighbors.	1.9820	0.6659
I have	17	I have clear expectations of what my community can achieve.	1.6602	0.6750
I can	18	I take the necessary steps to achieve my objectives.	1.6287	0.6501
I can	19	I can use humor as a way to cope.	1.7919	0.7569
I am	20	I am good at my work.	1.8982	0.6111
I am	21	I am proud of myself, my family, and my community.	1.5853	0.7648
I can	22	I can see several sides of a situation.	1.7650	0.6595
I can	23	I can take care of myself, and my family.	1.8009	0.6462
I am	24	I am considerate of the viewpoints of others in difficult situations.	2.1826	0.6988
I have	25	I have people in the community who listen to my worries.	2.0314	0,6390

Reliability coefficients (25 items) alpha=0.9365.

REFERENCES

Bandura, A. Human Agencies in Social Cognitive Theory. *Am. Psychol.* **1989,** *44*(9), 1175–1184.

Bernard, B. *Fostering Resilience in Kids: Protective Factors in the Family, School and Community;* Western Center Drug-Free Schools and Communities: Portland, OR, 1991.

Blum, R. M.; Rinehard, P. M. *Reducing the Risk: Connections that Make a Difference in the Lives of Youth;* Add Health Project: Bethesda, MD, 1997.

Bogenschneider, K.; Small, S.; Riley, D. *An Ecological, Risk-focused Approach for Addressing Youth-at-risk.* National 4-H Center: Chevy Chase, MD, 1993.

Grotberg, E. H. *A Guide to Promoting Resiliency in Children: Strengthening the Human Spirit;* Bernard Van Leer foundation: Amsterdam, The Netherlands, 1995.

Kass, J. D. The Inventory of Positive Psychological Attitudes. In *Evaluating Stress. A book of resources;* Zalaquett, C. P., Wood, R. J., Eds.; Lanham, Md, and London: Scarecrow Press, 1998; Vol 2, pp 153–184.

Luthar, S. The Construct of Resilience: A Critical Evaluation and Guidelines for Future Work. *J. Child Dev.* **2000,** *71*(3), 543–558.

Luthar, S.; Zigler, E. Vulnerability and Competence: A Review of Research on Resilience in Childhood. *Am. J. Orthopsychiatry* **1991,** *61*(1), 6–22.

McCubbin, H.; Marilyn, I.; McCubbin, A.; Thompson, A. I. Resiliency in Families: The Role of Family Schema and Appraisal in Family Adaptation to Crisis. In *Family Relations: Challenges for the Future;* Brubaker, T. H. Ed.; Sage: Beverly Hills, CA, 1993.

Neill, J. T.; Dias, K. L. Adventure Education and Resilience: The Double-Edged Sword. *J. Adventure Educ. Outdoor Learn.* **2001,** *1*(2), 35–42.

Rutter, M. Psychosocial Resilience and Protective Mechanisms. *Am. J. Orthopsychiatry* **1987,** *57*(3), 316–331.

Tedeshi, R.; Park, C.; Calhoun, L.; Eds. *Postraumatic Growth: Positive Changes in the Aftermath of Crisis;* Erlbaum: Mahwah, NJ, 1998.

Valliant, G. E. Mental Health. *Am. J. Psychiatry* **2003,** *160,* 1373–1384.

Wagnild, G. M.; Young, H. M. Development and Psychometric Evaluation of the Resilience Scale. *J. Nurs. Meas.* **1993,** *1,* 165–178.

Werner, E. E.; Smith, R. S. *Overcoming the Odds;* Cornell University Press: Ithaca, NY, 1992.

CHAPTER 6

A HISTORICAL OVERVIEW OF PSYCHOSOCIAL SUPPORT PROGRAM RESPONSE TO THE 2004 SOUTH ASIA TSUNAMI[1]

ANJANA DAYAL De PREWITT[2]

Associate Director, Psychosocial Program, Save the Children USA

CONTENTS

---

[1] The American Red Cross was operating a psychosocial support program in India as part of the Gujarat Earthquake 2001 long-term recovery. By 2004, the program was transitioning and a process of reengineering was taking place. The December 25, 2004 tsunami precipitated a recall of the psychosocial support team back to full operations.

[2] Ms. Dayal de Prewitt was the architect of the immediate response in conjunction with Erick Baranik, and James Randy Ackley.

6.1   INTRODUCTION

This chapter provides a description of the emergency response to the South Asia tsunami, 2004. This chapter is divided into three parts. The first part presents a brief review of the literature. The second part is a description of the American Red Cross (ARC) psychosocial response to the tsunami during the emergency phase. The third part presents a case study general overview of the development of the long-term psychosocial support intervention in Sri Lanka.

One of the most powerful earthquakes on the planet hit the Asian continent with its epicenter at Aceh, Indonesia. Some of the tsunamis reached as far as 1650 km (1200 miles) from the epicenter of the 9.0 magnitude quake, which was located about 160 km (100 miles) off the coast of Indonesia's Sumatra Island at a depth of about 10 km (6.2 miles). (www.lk.undp.org).

The Indian Ocean tsunami disaster resulted in one of the largest relief, recovery, and reconstruction operations ever launched by the International Federation of the Red Cross and Red Crescent movement (IFRC). For the ARC, international services, and psychosocial support programs (PSP), a relatively new addition to its emergency response repertoire became a program of great importance and assistance to the survivors of this terrible disaster. This intervention changed the nature of PSP during emergency responses in years to come (IASC 2007, IASC 2012). This chapter presents and discusses the immediate response and the plans of long-term implementation of psychosocial support in the Republic of Maldives and Sri Lanka.

6.2   IMPACT OF DISASTERS AND LOSS OF PLACE

Adverse events, such as disasters, affect communities and can produce such human and material losses that the resources of the community are overwhelmed, and therefore, the usual social mechanisms to cope with emergencies are insufficient (Lopez Ibor, 2005) sugests that a disaster causes a total breakdown of everyday functioning, the normal social functioning disappears, there is a loss of the trusted elderly and the community leadership, and the health and emergency systems are overwhelmed in a way that survivors do not know how to receive help.

Lazarus (1966) defined the disaster-related stress in terms of the transaction between the person and their environment ("sense of place"). Barton (1969) suggests that disasters challenge the social sustainability of the affected person by altering the nature and quantity of "inputs" to their community, thereby producing changes in the "demands" from the perceived "place." The demand-adaptation model proposed by Barton (1969) has been studied widely in disaster studies focusing on the elements of "place" including (1) individual (Fritz and Marks, 1954), (2) families (Bolin, 1976), (3) organizations (Dynes, 1970; Perry et al., (1974), (4) communities (Hass et al., 1977), and (5) societies (Lessa, 1964). A disaster causes survivors the actual or perceived loss of social networks, governmental systems, and individual indicators of control that lead to individual and community psychosocial competence.

The tsunami disrupted the affected peoples' sense of place. People were dislocated from their communities to shelters and camps. Their psychosocial network is disrupted and the feelings of safety and emotional comfort are disrupted. Affected people experienced loss of community commitment, thus, exhibiting reduced desire to participate and volunteer in community activities. Affected people and communities experience a social sustainability challenge.

Loss of place causes disruption of the collective sense of safety and destroys lifestyles, lives, and places. This disruption brings about a rupture in the emotional bonds that people form over time, the loss of strongly felt values, meanings, and symbols, the quality of place that people may not be consciously aware of until it is threatened or lost. It destroys the set of places' meanings that are actively and consciously constructed and reconstructed within the affected people minds, shared cultures, and social practices, and the geographic and spatial context within which meaning, values, spiritual and psychosocial interaction occur (Prewitt Diaz and Dayal, 2008).

Disaster-affected people develop a sense of being out of control and experience helplessness and vulnerability. Trauma causes people to see the world defiantly so some will be drawn together in groups to help reestablish boundaries and structures—a new sense of place. Others feel estranged and want to be alone. Traumatic events can mobilize and strengthen or it can fracture a family, group, or a community.

Disasters break or fracture the social and psychological community networks instantly (the displacement and relocation of thousands after Katrina in 2005). When the physical, social, or psychological place is destroyed they must be grieved in ways like mourning a death. Losing access to places of cultural and psychological, spiritual or social significance, and the resulting loss of connection to other affected people undermine a community's ability to act causing grieving to occur.

These psychosocial reactions are problematic both in the immediate aftermath of the event and during rehabilitation and reconstruction; disasters hinder a persons' ability to integrate their past and present life. The person is impaired without the help of tangible social and environmental cues and symbols (Brown and Perkins, 1992). Places that were connected to them become images and memories. These events set a chain of events that triggers further events and response which may last several years. For some disaster-affected persons' the impact of a disaster may last several years or perhaps a lifetime (Perry and Lindell, 1978).

Perry and Lindell (1978) conclude that disasters cause changes in the psychosocial, environmental, and spiritual lives of the disaster-affected people which in turn cause the perception of "place" to change and adapt to different demands. Disaster-related stress occurs when there is an alteration of the perception of "place" and the resulting challenges: loss of community commitment, participation, and care for the natural environment.

It may take months before the persons affected by an adverse event can reestablish their sense of place. The persons begin to feel comfortable in their new environment, the systems begin to work once again (health and welfare are reestablished, schools are back in sessions, livelihood activities have been initiated), and the social networks become functional, too. Once the person begins to function normally within a community, it is said that resilience is enhanced and the survivors become stronger and realize that they have the capacity to survive.

6.3   CONSIDERATIONS IN FORMULATING THE PSP RESPONSE TO THE TSUNAMI

The ARC PSP response was driven by a request from the IFRC Societies' field assessment and coordination team (FACT) in the Republic of Maldives. The IFRC realized that there were many people experiencing disaster-related distress and that the country did not have the human resources needed to alleviate this stress in schools or communities on this island. In many of the islands, the helpers were also survivors themselves. In consultation with the FACT Team in the Maldives, ARC personnel in New Delhi, India prepared a concept paper that described two objectives for the PSP intervention: (1) as psychological first aid training to 57 counselors and (2) equipping 328 teachers (the total teacher workforce at the time) with the necessary tools to alleviate disaster-related stress as well as implementing self-care activities for all the children.

6.4   THE EMERGENCY PHASE IN THE REPUBLIC OF MALDIVES

A two-person team consisting of a psychiatrist and a social worker arrived in the Republic of Maldives on January 3, 2005. The objective was to conduct a 2-day psychological first aid course and to prepare personnel to go out to the affected atolls, provide psychological first aid, and conduct assessments. In the meantime, the ARC team conducted workshops with counselors to develop skills in psychological first aid and crisis intervention, thus building the capacity of the Department of Mental Health (DMH) to implement community-based interventions.

Once the teams returned from the atolls with data on the emotional and practical needs, the FACT-IFRC team, the ARC PSP team, and the Ministry of Health of the Republic of Maldives developed a three-tier program focusing on community-based intervention models. They identified the urgent needs to train teachers, nurses, and representatives from health and education sectors, and provide direct services to the affected population in the Republic of Maldives.

TABLE 6.1   Primary and Secondary Responsibilities for the Maldivian Personnel Trained in Psychosocial First Aid and Sent to the Affected Atolls.

Primary responsibilities	Secondary responsibilities
Provide support, care, and understanding to person exhibiting signs of distress	Mobilize and advocate staying together by strengthening community networks
Accept information and feelings of the persons	Keep the family groups together.
Provide information:	Make sure that children are supported and kept active
• Determine what has happened and what will happen	• Help to facilitate and perform rituals
• Provide the telephone for the counseling center	• Support on-the-scene visits
• Help facilitate contact with other relatives	• Support spaces for memorial activities
• Provide clear information about the normal reactions to this abnormal situation	• Ensure follow-up over time

A second team, consisting of one psychologist and one community developer, departed New Delhi on January 10, 2005, with the objective of developing a culturally/linguistically and contextualized teacher manual and training six teachers in the Maldives to share the manual's contents with their peers. These teachers became the trainers and multiplied the two trainings throughout the atolls. The ARC/PSP team met for 3 days with the Maldivian teachers and developed three chapters that included some theory and a lot of practical exercises. Three teams of three persons, (two Maldivian teachers and a PSP staff) participated in a massive campaign to develop the psychosocial capacity of all teachers in the Republic of Maldives. During the remainder of January and 3 weeks in February, the teams traveled all the islands from the Indian Ocean to the Arabian Sea. At the end of this period, 328 teachers had participated in the capacity-building sections.

By the third week of February 2005, the "Training of Trainers" workshop had enhanced all the Maldivian teachers to take over the implementation of the psychosocial program. Each school received a school chest from ARC consisting of school materials and games. When the schools opened, in the latter part of February, the teachers could respond effectively to the children's disaster-related stress. The initial response was completed in 7 weeks and handed over to the Republic of Maldives[1].

6.5   THE EMERGENCY PHASE IN THE DEMOCRATIC SOCIALIST REPUBLIC OF SRI LANKA

The Sri Lanka psychosocial response was the largest and most complex of the three programs operated by the ARC as part of the tsunami response (Ring, 2005). Early conversations in the third week of January between ARC, Sri Lanka Red Cross Society (SLRCS), the IFRC, and other stakeholders opened the doors to the ARC PSP to initiate operations in the southern and western provinces. It was agreed that the Danish Red Cross would continue to operate under the umbrella of the International Committee of the Red Cross (ICRC) in the north and east. By the 3rd week of February, a team composed of a social worker and a psychologist came to Sri Lanka to conduct a detailed need assessment in the southern and western provinces.

The team, based in Galle, began to analyze secondary sources from the region, train Sri Lankan Red Cross volunteers in psychological first aid and rapid needs assessment skills, visit schools and communities, and assist the Federation's emergency relief unit (ERU) to attend to the emotional needs and manifestations of disaster-related distress during relief distributions. After the emergency phase (8 weeks), the team concluded, based on the community assessments, that many more persons were needed to conduct an in-depth assessment, psychological first aid and sense of place activities in the affected communities.

By mid-March, a team of four ARC/PSP personnel was assigned to the IFRC as staff-on-loan for a period of 3 months. The team's objective was to utilize a participatory approach to assess community and school needs, prioritize the psychosocial needs, and develop with the primary stakeholders (the SNRC and the affected people) psychosocial support interventions that would be agreed upon by the target community and schools. Once the assessment was completed, the staff developed concept

---

[1] Until 2010, the ARC community-based psychosocial support was maintained in the Republic of Maldives supporting predominantly community mobilization and reestablishment of place activities.

papers and began to write a PSP proposal with the help of the technical assistance unit of the International Services of the ARC (Ring et al., 2005; Prewitt Diaz, 2012).

All ARC/PSP personnel were pulled from their assigned branches in the southern and western provinces from June to August 2005, until the proposal was evaluated and funding secured. ARC/PSP personnel returned to the branches in August 2005 to begin the long-term recovery activities. They were met with frustration, despair, and disbelief. It would take approximately the next 3 months to build back the trust and support of the community to continue and organize them to take care of each other's emotional and social needs. In February 2006, a year after the tsunami had impacted the Spillanian coast, the SLRCS/ARC PSP could hire their first local staff members in Sri Lanka.

6.6 LESSONS LEARNED DURING THE EMERGENCY PHASE

The first lesson learned from these early weeks of the response was the inefficiency of placing the base of operations for the PSP tsunami response in New Delhi when the actual work was taking place in the Republic of Maldives. All logistics were handled from New Delhi including recruitment, briefing, communications, funding, assembling of school chests and transporting them via air to the Maldives. There were challenges in this cumbersome setup: (1) The government of India had not given ARC permission to operate outside of India; (2) the Indian Red Cross Society (IRCS) denied ARC permission to send goods and personnel outside of India; and (3) The ARC delegation, with a focus on long-term programming, was not prepared to run disaster operations at the regional level. In the future, more attention should be given to modifying a country office into a disaster operations center.

The second lesson was the response of the primary stakeholders once the ARC/PSP staff members were pulled out to develop concept paper and proposals. This gap in service had detrimental effects on the primary stakeholders, it generated feelings of abandonment. The transition between the rehabilitation phase and long-term recovery resulted in a "second disaster" for the primary stakeholders. Mechanisms should be developed that ensures the continuity of services from one phase of the disaster response to the next.

6.7   ONWARD TO LONG-TERM DEVELOPMENT PROGRAMS: CASE STUDY OF SRI LANKA

The tsunami of December 26, 2004, displaced almost 1 million people and killed more than 30,000 people in Sri Lanka. This was an unprecedented disaster in recent Sri Lankan history and posed many challenges for the reconstruction and rebuilding of people's lives. The social and psychological loss could not be measured: the loss of loved ones, the shattered lives and lifestyles, and people left without identity, autonomy, and what psychologists term a "sense of place." (Fullilove, 1996, 2004) The tsunami's impact was far-reaching for the survivors' psychosocial health, and well-being had been affected in such a way that "root shock" (Fullilove 2011) was readily observable; steps had to be taken to address the situation both immediately and over the long term. The project was finally transferred to the Sri Lanka Red Cross in 2010 (Final Evaluation Report, 2011).

The ARC supported PSP long-term psychosocial response was informed by guidelines proposed by the Ministry of Health of Sri Lanka (January 10, 2005), the Consortium for Humanitarian Agencies (2005), the SPHERE project (2004) as well as the Centre for National Operations (CNO) (2005). The CNO articulated three objectives for long-term reconstruction activities in psychosocial support: (1) continue to support trainings and other psychosocial activities, (2) develop further infrastructure for psychosocial support in Sri Lanka, and (3) help to procure funding for suitable organizations to carry out their work on health and well-being. The publication of the Inter-Agency Standing Committee/Mental Health and Psychological Support (IASC/MHPSS) guidelines (2007) not only continued to inform the program but also are provided a roadmap for long-term development.

The PSP proposal was submitted to the ARC, the international federation, and the Sri Lanka Red Cross. The psychosocial technical committee approved the proposal. Positions were advertised, and hiring was completed by mid-December 2006, almost 1 year after the tsunami. The first group of Sri Lanka Red Cross personnel participated in capacity-building activities at the beginning of January.

The PSP was based on five strategies: (1) using a community-based approach, (2) making sure that interventions are contextual and culturally and linguistically appropriate, (3) empowering affected people, (4) encouraging community participation, and (5) encouraging active involvement.

6.8   COMMUNITY-BASED APPROACH

Experience has shown that when implementing PSPs, community-based approaches are best (Prewitt Diaz et al., 2004a). Building on local resources, providing training, and upgrading local structures and institutions are critical to the success of a psychological support program and would reestablish place (Fullilove, 2004). This approach allowed trained volunteers to share their knowledge with fellow community members. Because many emotions (e.g., distress and sorrow) do not require professional treatment, these local resources often become instrumental in providing successful relief (Project SPHERE, 2004). A much larger number of people can be helped by working with groups rather than individuals and focusing on strengthening networks in the community. In addition, involving the community with its knowledge, values, and practices makes a culturally appropriate response more liable (Prewitt Diaz et al., 2004b).

6.8.1   INTERVENTIONS ARE CONTEXTUAL, CULTURALLY AND LINGUISTICALLY APPROPRIATE

The international arena where most ARC/PSP programs are implemented presents the challenge of diverse languages, cultures, and religious beliefs that are not frequently understood by an outsider and is more of a hindrance than a help during the immediate aftermath of an adverse event, such as the tsunami. Culture exercises a great impact on the way in which people view the world. Building the capacity of community volunteers ensures inside knowledge of the local culture and utilizes their ability to provide culturally appropriate assistance to the affected population (Adler, 1991). Trained personnel from the disaster-affected community can react immediately in times of crisis and can assist with the provision of long-term support to the survivors. They have easy access to, and the confidence of, the disaster survivors.

6.8.2   EMPOWERMENT

Accepting help may be the beginning step as part of the positive process and may solve a crisis for the affected individual. However, to the survivor, it may also emphasize inability and dependency, leading to bitterness or anger about being a victim in the eyes of others. The ARC PSP program knows high-quality psychosocial assistance and is based on helping others to regain self-respect and autonomy. It puts the emphasis on the abilities and strengths of recipients more than on their problems and weaknesses. A high degree of community participation is generally accepted as an effective way to encourage empowerment of the people (International Federation of Red Cross and Red Crescent Societies, 2003).

6.8.3   COMMUNITY PARTICIPATION

Basing projects on ideas developed by concerned people themselves will promote empowerment and local ownership and help facilitate and consolidate a long-term capacity for problem-solving. Through participation, people gain an increase in control over their lives as well as over the life of the community. Participation in collective decision-making about their needs, as well as in the development and implementation of strategies, is based on their collective strength to meet those needs.

6.8.4   ACTIVE INVOLVEMENT

The ARC PSP focuses on individual strengths and provides spaces so that community members can enhance their resilience by building on existing resources, coping mechanisms and resiliencies. The objectives of the interventions are: (1) identifying and strengthening internal coping mechanisms, (2) fostering active involvement of people in community mapping and identifying problems and resources, and (3) recognizing people's skills and competence.

Self-help actions and strategies adopted by the affected populations themselves are a key to their successful recovery. To listen to the communities' language of distress and to identify the community activities to help alleviate such distress is crucial in the planning of interventions. In major adverse events, community and family support structures may have broken, and they should be rebuilt. The focus of the PSP is on people's positive efforts to deal with and come to terms with their experiences.

Since the beginning of the program, 6 months have passed. There is more than one hundred SLRCS psychosocial staff in seven sites, supported by four delegates in all major communities in the southern and western provinces. To date, the PSP has provided direct services to 46,751 persons in the service area.

The program is currently focusing on four programmatic areas. First, there are capacity-building activities at several levels. All SLRCS project staff received capacity building at the outset of their contracts. As of now, there are 76 SLRCS personnel trained as crisis intervention specialists. Teachers and community facilitators are participating in staff development activities. So far, 91 teachers and 290 community facilitators have completed their basic 36-hour capacity building. Thirty-one preservice teachers are receiving a 1-week course at the Ruhenu Teacher Training College, which brings the total of preservice teachers trained to 152. Operational training on PSP and psychological first aid for volunteers and chapter personnel has taken place in all seven districts. Seventeen master trainers received instruction in a 1-week course on community-based psychosocial support by an external expert.

In addition, the PSP has two materials development centers one in Kalutara and the other in Matara. Both material development centers are working to produce trifolds, posters, manuals, and storybooks that are culturally, linguistically, and contextually appropriate. These centers are filling a gap by providing appropriate psychosocial materials in the Sinhalese language and culture.

The PSP is actively engaged in 58 communities of the target geographical area. Sixty-seven resilience activities (including sports activities, cultural events, and network-enhancing activities and community cleanups) have been implemented, and six community centers have been reestablished. In the 6 months of the program, four resilience projects have been developed. These activities are somewhat complicated as they require the community to engage in mapping exercises, identification of existing risks factors, planning, developing, and evaluating a community project.

There are 51 schools that form part of the PSP activities in the seven districts. Until now, activities such as school mapping, painting of chairs, creative and expressive activities, school kits distribution, mural and wall painting, and sports events have taken place. A total of 28,556 children, teachers, and parents have been directly exposed to and positively affected by the PSP.

The PSP has established formal relations with the Ministry of Education and the Ministry of Social Services and Welfare. These relationships have led to participating in committees, conducting capacity-building activities, planning joint activities, and preparing a crisis intervention professional program of study. These interactions with the Government of Sri Lanka will help ensure the program's long-term sustainability.

6.9   CHALLENGES AND LESSONS LEARNED

Psychosocial support is a relatively new concept in the repertoire of immediate disaster response for the movement partners. The initial reaction from the federation and others is an acceptance that PSP should be a part of the immediate response. The evidence and best practices are inconclusive about the effectiveness of PSP during the reconstruction phase.

Language becomes the way in which survivors express distress and assign an emotional value to the communication. PSP personnel must understand the context of the overall community's mode of expression. Materials' development must be contextual, translation of the materials to the target language is not sufficient. A local person must, then, be asked to assess the materials and to make the content (language and pictures) fit into the local society and culture.

Many people utilize traditional healers and other respected members of the community. It is important to develop the capacity of the traditional healers and others so that they can provide psychological first aid and refer the affected member of the community to other sources as needed.

6.10   EVALUATION OF PSP DURING THE IMMEDIATE TSUNAMI RESPONSE

The federation commissioned a study on the effectiveness of its interventions and those of the local Red Cross and participating national societies (PNSs) during the immediate response and rehabilitation phase. Many findings from the study suggested that the response of the Red Cross movement had been less than effective. However, one of the key findings was the valuable role of psychosocial support during the emergency phase as well as the reconstruction phase of the tsunami Bhattacharjee (2005).

In Sri Lanka, Indonesia, and the Maldives, psychosocial work carried out by the PNSs has played a crucial role in the recovery process, addressing the massive psychosocial impact of the disaster-affected people living in shelters and temporary accommodations (p. 21). The report further indicates "in the Maldives, a PNS-run psychosocial program (ARC) continues to play a critical role in providing support to the communities" (p. 22).

The report suggests that the Red Cross partners exhibited an "obsessive preoccupation in Sri Lanka with adherence to bureaucratic procedures that have led to near-total disengagement with the affected communities." The report once again commends the PSP interventions. "Only in sectors like psychosocial support, [is] there a continuing engagement with the communities." (p. 36). In Sri Lanka, the ARC PSP seems to be experiencing success in achieving community ownership of the projects resulting from community-based planning, design, and implementation.

REFERENCES

Adler, N. J. *International Dimensions of Organizational Behavior,* 2nd ed.; PWS-KENT Publishing Company: Boston, MA, 1991; pp 63–91.

Bhattacharjee, A. *Real Time Evaluation of Tsunami Response in Asia and East Africa, Second Round;* Final Report. International Federation of the Red Cross and Red Crecent: Geneva, Switzerland, 2005.

Center for National Operations (CNO). *Description of the Psychosocial Support Desk in the CNO;* Center for National Operations: Colombo, Sri Lanka, 2005.

Inter-Agency Standing Committee (IASC). *Mental Health and Psychosocial Support in Emergency Settings;* IASC Task Force on Mental Health and Psychosocial Support: Geneva, Switzerland, 2007.

International Federation of Red Cross and Red Crescent. *Community-Based Psychological Support;* International Federation of the Red Cross and Red Crescent: Geneva, 2003.

Ministry of Health. *National Plan of Action for the Management and Delivery of Psychosocial and Mental Health Services for People Affected by the Tsunami Disaster;* Directorate of Mental Health Services: Colombo, Sri Lanka, 2005.

Patel, V.; Mutambirwa, J.; Nhiwatiwa, S. Stressed, Depressed or Bewitched. In *Development and Culture*; Eade, D., Ed.; Information Press (OXFAM): London, England, 2002.

Prewitt Diaz, J. O. *Operational Teacher Training for Tsunami-affected Area;* American Red Cross: Male, Republic of Maldives, 2005.

Prewitt Diaz, J. O.; Bordoloi, S.; Rajesh, A.; Sen Dave, A.; Khoja, A. *The Crisis Intervention Training Program;* Indian Red Cross Society/American Red Cross: New Delhi, India, 2003.

Prewitt Diaz, J. O.; Bordoloi, S.; Mishra, S.; Rajesh, A.; Dash, S.; Mishra, I. *School-based Psychosocial Support Training for Teachers;* Indian Red Cross Society/American Red Cross: New Delhi, India, 2004a.

Prewitt Diaz, J. O.; Bordoloi, S.; Mewari, M. *Crisis to Recovery: The Road to Resiliency;* Indian Red Cross Society/American Red Cross: New Delhi, India, 2004b.

CHAPTER 7

COMMUNITY-BASED PSYCHOSOCIAL SUPPORT IS A PROCESS AND TOOL FOR PROTECTION OF THE VULNERABLE SURVIVORS OF DISASTER

SUBHASIS BHADRA

Associate Professor, Department of Social Work, School of Social Sciences, Central University of Rajasthan, Bandarsindri, NH-8, Dist. Ajmer, Kishangarh, Rajasthan 305817, India

CONTENTS

7.1 INTRODUCTION

"Human rights" is the motto of civilization of the modern world. Human Rights declarations, laws and directives strives to alleviate people's undue suffering. The first humanitarian movement of Henry Dunant through the birth of Red Cross established the rights of the wounded sick soldier (ICRC, 2008). In due course, over two World Wars the League of Nations developed and evolved as United Nations (UN), and finally the "Universal Declaration Human Rights (UDHR; 1948)" established the true spirit of civilization. Protection is an important component of human rights. All humans are born free and equal in dignity and rights. Every human being should live in the frame of justice and peace which are inalienable and inherent within the ethos of UDHR. Human rights are available to all to ensure the enjoyment of decent life in any condition. Disaster is one of the most life-threatening experiences for the survivors as it causes multiple impacts, and extensions of damage continue to impose various hardships in the life of the survivors during, immediately after the disaster, and in long-term. Human rights of the survivors largely get affected if the existing federal administration fails to take adequate measures and ensure safety and security at right time. There is an inherent tendency of the post-disaster situation, characterized by chaos to keep the system disturbed if the pre-disaster preparedness and post-disaster interventions are ineffective.

Human rights are considered as the valid natural justifiable claim of the human being. These are universally applicable and true for every human being. Every claim of right is inconsistent with an owed duty. A claim of right is held by someone means there must be another person/s or institution, or system that owes a corresponding duty. The effective functioning of the duty bearer is important for fulfilling the human rights claim. Some important dimensions of human rights are equality, freedom, protection, justice, and dignity. In the context of disaster, all these dimensions are severely shaken up and for the weaker, marginalized section, and the pre-disaster vulnerabilities become even more critical and deep-rooted. There are different policy guidelines, acts, and mechanism at the International level and in India to deal with disasters interventions and to ensure an effective outcome in the phase of rescue, relief, rehabilitation, and in long-term recovery.

Psychosocial support (PSS) is considered as one of the crosscutting interventions strategies in different guidelines (Inter-Agency Standing

Committee (IASC), 2007; the Inter-Agency Network for Education in Emergencies (INEE), 2005; The SPHERE project, 2011) that has been practiced universally by many international organizations, such as ICRC, International Federation of Red Cross (IFRC), American Red Cross (ARC), International Medical Crop, Medicine Sans Frontier, United Nations International Children's Emergency Fund (UNICEF), United Nations Development Programme (UNDP), and so forth. Community-based psychosocial support (CBPSS) is an important model in this direction to facilitate recovery and ensure well-being of the survivors. Disaster is defined by its impact on human being or survivors. An event impacting human life becomes a disaster, thus water upsurge in Barren Island is not flooded, or shaking of the earth in the desert is not an earthquake that drives any human attention. The gravity of impact on survivors is crucial for understanding the severe nature of the disaster, but all the survivors do not get affected equally based on their multiple sociodemographic features. There are few sections of the population that are generally considered as vulnerable who bear the brunt of disaster more than others and need consistent considerable intervention to recover and regain normalcy. Thus, human rights, protection and PSS in the community belong in the same continuum of disaster recovery explained in this chapter.

7.2   HUMAN RIGHTS IN INDIA AND WORLDWIDE ABOUT SURVIVORS OF DISASTER

The first Geneva Convention of 1864 is an important milestone in the development of human rights. The founder of Red Cross Movement, Henry Dunant, was the main person for paving the way for humanitarian concern for care of the wounded sick military personnel without discrimination (United for Human Rights, 2016). In 1945, the UN Charter was adopted that established the fundamental pillars for international peace and security, international development and cooperation, protection and respect for human rights and the rule of law. In 1948, UDHR was considered as decisive landmark in the struggle for establishing human dignity. Fifteen years later the General Secretary of UN, Mr. Javier Perez de Cuellar mentioned that UDHR has become the basic international code of conduct by which performance in promoting and protecting human rights is to be measured. The Declaration does not establish any difference in value or importance between economic, social, and cultural rights, on the one hand, and civil and political liberties, on the other. They are granted the same degree of protection (Sekar, 2008, p. 246).

There are some important UN conventions that established the specific needs of protection for the vulnerable sections and people in general (UNOHCHR, 2016).

1. International Convention on the Elimination of All Forms of Racial Discrimination, 21 Dec 1965
2. Convention on the Elimination of All Forms of Discrimination against Women,18 Dec 1979
3. Convention on the Rights of the Child, 20 Nov 1989
4. Convention on the Rights of Persons with Disabilities, 13 December 2006

Around the world, different countries have developed their own in-country mechanisms to deal with human rights issues and to ensure protection, freedom, and justice. Some important global achievements are the establishment of European Court of Human Rights (ECHR), American Convention on Human Rights 1978, African Charter on Human and Peoples' Rights 1986, and National Human Rights Commission (NHRC) in India 1993.

The United Nations Office for Disaster Risk Reduction (UNISDR) was established in 1999 to facilitate an international strategy for disaster risk reduction and made the considerable focus on protection of different human rights for the survivors of the disaster and for the vulnerable survivors. The major milestone was adopting The Hyogo Framework for Action (2005–2015): Building the Resilience of Nations and Communities to Disasters in 2005. Fourth Asian Ministerial Conference on disaster risk reduction, held in Korea, in October 2010 specifically recognized "the need to protect women, children, and other vulnerable groups from the disproportionate impacts of disaster and to empower them to promote resiliency within their communities and workplaces" (UNISDR, 2010).

Though many countries have initiated various mechanisms to protect and promote human rights; yet after almost seven decades of the UDHR, the situation of human rights is under criticism with increasing human sufferings all over the globe. Beyond poverty and disease, natural and human-made disasters, complex emergencies also lead to huge refugee, and displacement imposing challenges to protect human rights of the survivors.

In India, human rights mechanisms are established and yet to evolve and be effective in all the critical areas, and in disaster issues. The NHRC, is the key responsible body to establish human rights in India, was established on 12th October 1993 under the legislative mandate of the Protection of Human Rights Act, 1993. Within the purview of NHRC, various aspects of civil and political, as well as economic, social, and cultural rights are included. From time to time NHRC focused on specific areas of human rights concern in the matter of terrorism and insurgency, custodial death, rape and torture, and issues of human rights of the survivors of disasters. In different incidents of disasters in India, the NHRC took proactive steps in formulating various directives to the states, ministries, and departments about human rights protection and ensure social justice against any violation of rights of the survivors. The protection of the human rights is a very intricate issue within the dynamics of the community, social, administrative systems that can only be assured by community engagement and community participation in the process of disaster intervention, in pre-, during, and post-disaster phases. Thus, protection of human rights can effectively be delivered and ensured through community-based mechanisms and CBPSS.

7.3 VULNERABLE SURVIVORS OF DISASTER AND PROTECTION ISSUES

Disaster survivors need protections as the event of disaster expose them to a dilapidated psychosocial condition, and breakdown of the social-ecological balance require a considerable amount of time and effort. Thus, protection of human rights is a basic deliberation in any circumstance both in natural or human-made disasters. Within the frame of PSS human rights protection is most crucial for the disaster intervention. In the SPHERE project protection principles are considered as the core guiding value, as well as, the IASC standards (The SPHERE project, 2004; The SPHERE project, 2011) are developed keeping the human rights at the center. "Protection and Human Rights Standards" is a common function as per the IASC-MHPSS Guidelines (IASC, 2007) during emergencies. The constitution of India also upholds human rights principles and the issues of protection for its citizen. The vulnerable and marginalized sections face more difficulties and human rights violations, calls for active protective measures and actions.

The SPHERE project (2004, 2011) mentioned the factors of vulnerabilities and issues of marginalization are closely interlinked. Many vulnerable groups have been specified who tend to get marginalized and neglected in disaster response. "Failure to recognize the differing needs of vulnerable groups and the barriers they face in gaining equal access to appropriate services and support can result in them being further marginalized, or even denied vital assistance" (SPHERE, 2004, p. 9). Twigg (2004) in "Good Practice Review" pointed out "certain groups are particularly vulnerable to disaster: they include people marginalized by gender, age, ethnicity, and disability. The root causes of their vulnerability lie in their position in the society." Occasionally, marginalization is a deliberate action by the system (in the community of authority) but innate vulnerabilities render these groups less capable to reach out for assistance (Barber, 2015; Bhadra, 2016). The important vulnerable groups are women (single, widowed, injured, and women with young children), children (orphan, uncared, and young), disabled and injured, older people (uncared and sick), and people living with HIV/AIDS, who usually tend to get marginalized.

They become the double victim of a disaster as their subjugated social status, complications in a post-disaster scenario cause considerable disabling circumstances. The severity and impact of disaster cause breaking down of the social systems and the prevailing disorganized condition fail to ensure any social support for the vulnerable population that might have been existed before the disaster or fails to create a new mechanism of support immediately after the disaster. Thus, the vulnerable survivors get into a vicious circle of sufferings with complicated, undesirable, and unpredictable life events. These life events lead to issues, such as human trafficking, forced migration, child labor, loss of nurturing environment, and fragile social institutions. Every disaster in India has caused huge trafficking for the unprotected adolescent girls, children (UNICEF, 2012; Bhadra, 2015), similarly, caused migration out of poverty to search for livelihood (Companion, 2015) and huge sufferings for the disabled (Kelman and Stough, 2015; Bhadra, 2015).

The factors of vulnerability and corresponding marginalization vary as per the nature, time, space, and situation of the disaster. Very commonly a developed country reaches out to the survivors quickly and facilitates rebuilding at a faster rate and reduces the vulnerabilities than the developing or underdeveloped countries. In such situation, the survivors continue to live with an increased number of vulnerability factors. "People in countries ranked among the lowest 20% in the Human Development Index are 10–1000 times more likely to die in a natural disaster than people from countries in the top 20%" (Leidig and Teeuw, 2015). Protection is a central theme of human rights and IASC guidelines on Protecting Persons Affected by Natural Disaster mentioned (IASC, 2006a) "protection is not limited to securing the survival and physical security of those affected by natural disasters. It encompasses all relevant guarantees—civil and political as well as economic, social, and cultural rights." In the account of the wider impact of any disaster, multiple issues of protection for the vulnerable sections start surfacing that relate to four basic dimensions of human rights.

1. Rights related to physical security and integrity:
   a) Sexual assault of the children, adolescents, girls, and women
   b) Physical assault, torture, and violence
   c) Deliberate harm, forced labor, and displacement
   d) Harm to psychological integrity and dignity
2. Rights related to basic needs:
   a) Inadequate food, clean water, and sanitation health care
   b) Inadequate shelter, lack of privacy (specifically for adolescent girls), break up of family
   c) Problem of reproductive health care (specifically for adolescent girls)
3. Rights related to other economic, social, and cultural protection needs
   a) Denial of the right to education, healthcare, and adequate shelter/housing
   b) Discriminatory practice against children, adolescents, women, and persons with disability
   c) Loss of property, belonging, livelihood, and survival strategies
4. Rights related to other civil and political protection needs:
   a) Separation from family members
   b) Denial of freedom of expression, speech, association, and religion
   c) Loss of certificates and personal documents
   d) Arbitrary restrictions on freedom of movement, including punitive curfews or roadblocks that prevent access services. Like educational, medical facility, and economic activity

7.4   GUIDELINES FOR PROTECTION OF VULNERABLE POPULATION IN DISASTERS

The SPHERE project is an important document to set the standard response strategies in different sectors (e.g., water and sanitation hygiene, shelter, healthcare, and food supply and nutrition) for the survivors of a disaster that highlighted four protection principles (The SPHERE Project, 2011, p. 4):

1. Avoid exposing people to further harm because of your actions
2. Ensure people's access to impartial assistance—in proportion to need and without discrimination
3. Protect people from physical and psychological harm arising from violence and coercion
4. Assist people to claim their rights, access available remedies, and recover from the effects of abuse.

These protection principles underline the human dignity, social justice, and respect after any incident of disaster for the survivors, who live in a critical socioeconomic-political context. The IASC-MHPSS guidelines mentioned "identify, monitor, prevent, and respond to protection threats and failures through" social protection, legal protection, and adopting human rights framework through mental health and PSS program. Prevention of gender-based violence is equally important in the post-disaster scenario as often lack of safety and security in the relief camp or temporary shelter fail to provide adequate privacy and women become target of every type of exploitation. The guidelines *Women, Girls, Boys and Men: Different Needs—Equal Opportunities* (IASC, 2006b), categorized the protection activities under three headings: responsive action; remedial action; and environment building. The general protections are to ensure humane treatment of the survivors, protection of civilian population, and prohibition of the specific use of weapons (Chemical and Biological); the specific protection encompasses protection against sexual violence, women deprived of liberty, preservation of family link, and protection of material health. A right-based participatory community approach is being advocated in the post-disaster scenario. The rights-based community approach for the vulnerable population is most appropriate that identifies right holders (women and children disabled) with their entitlements in one hand and duty bearers (local agencies, state, community, leaders, family, etc.) with their obligations on the other hand and seek to strengthen the system to claim and deliver. The INEE sketched the details of protection mechanisms and emphasized on psychological, physical health interventions, protection of the children in disaster, facilitating hygienic practices, promoting life skills education, working with the community, and participatory approach in the educational intervention (INEE, 2005). The Convention of the Rights of Persons with Disabilities (UN, 2007) passed by the United Nations Ad Hoc Committee on Disability in 2007 under Article 11 proposes that "State parties shall take, in accordance with their obligations under international law, including international humanitarian law and international human rights necessary measures to ensure protection and safety of persons with disabilities in situations of risk, including situations of armed conflict, humanitarian emergencies and the occurrence of natural disasters" (UN, 2007). The IASC Guideline on Humanitarian Action and Older Persons mentioned (2008) that often the geriatric population face social and economic marginalization and they are less protected in the situation of hazard and crisis. Ensuring care, social security, and engaging them in decision-making is very important (IASC, 2008).

The international guidelines, evolved through human rights declaration, humanitarian law, and concern over the human sufferings due to disasters and crisis, made various recommendations to promote the protection and create a conducive atmosphere for the vulnerable survivors. These common unique features of recommendations are as follows:

- Protection from any form of violence, exploitation, and inhuman treatment
- Community-based approach
- Participatory mechanisms at the grassroots levels
- Special care provisions for the marginalized and vulnerable
- Establishing mechanisms of care, support in the community
- Closely working with family members and empowering the families establishing family links
- Development of institutional care
- Establishing, strengthening, maintaining the government systems for effective policy and program development and implementation

The CBPSS program is universally accepted model that functions from the grassroots level to the higher-ups for streamlining the system in most effective and diligent manner in view of human rights principles and protection as the core feature with the efficient inclusive approach.

7.5 COMMUNITY-BASED PSYCHOSOCIAL SUPPORT AND RELATED CONCEPTS

Protection as the central theme of human rights and social justice principles are the driving force of the disaster interventions to response to the humanitarian crisis is well recognized, worldwide. The most effective tool for the same is described here as CBPSS which is not limited within the outcome of resiliency, rather inbuilt within the process of recovery through community engagement based on the strength of the community and adaptive capacity which enhance capability and well-being of the survivors. The roots of CBPSS can be traced within the dimensions of health and definition of mental health given by World Health Organization (WHO). The physical, mental, and social dimension of health was added with spiritual health for a broader perspective of reaffirming holistic well-being (Larson, 1998; Dhar et al., 2011). Mental health is a "state of well-being in which the individual realizes his or her own abilities, can cope with the normal stresses of life, can work productively and fruitfully, and is able to contribute to his or her community" (WHO, 2001). In accordance with these understanding, the PSS is a combination of psychological and social interventions (Hansen, 2008), a process of facilitating resilience within individuals, families, and communities (IFRC, 2005) and a broad range of community-based interventions that promote the restoration of social cohesion and dignity of individuals and groups (Aarts, 2000). Psychosocial support is "any type of local or outside support that aims to protect or promote psychosocial well-being and/or prevent or treat mental disorder" (IASC, 2007), and SPHERE handbook (2011) mentioned "some of the greatest sources of vulnerability and suffering in disasters arise from the complex emotional, social, physical, and spiritual effects of disasters" that need to be supported with a structured PSS program.

Psychosocial support is closely interlinked with the development of resiliency which is better conceptualized as ability or process rather than an outcome, and focusing on adaptive capacity at individual and community levels than recovery (Norris et al., 2008; Walker and Westley, 2011). Targeted intervention and community engagement are prime focuses of resiliency building that is built on the strength of the community, but it does not exclude the intervention by government or external agencies. The resiliency building is an empowering process considering the "strength-based perspective" (SBP) (Zastrow, 2010) of the affected community. Barker (2003) defined empowerment as "the process of helping individuals, families, groups, and communities to increase their personal, interpersonal, socioeconomic, and political strength and to develop influence toward improving their circumstances" (p. 142).

The principles of SBP explain, every individual, family, and community has strength; trauma, abuse, illness, and struggle can be injurious, but they also can be source of challenge and opportunity; there is innate capacity of the community (survivors) to visualize the change and bring better developmental opportunities for themselves; collaboration with the client to ensure an equal footage as a stakeholder in the process of intervention; and every environment is full of resources (Zastrow, 2010, p. 73). The CBPSS thrives on its practice considering the capacity of the individual, family, and community that is mobilized through active participation, and allow the communities to take a decision about the desired changes being an active stakeholder in the process through collaborative practice, and strive to make the best use of the available resources and opportunities.

The process-oriented engagement in CBPSS may not pinpoint to a single outcome practically, but it is undoubtedly bringing a capability to the survivors that can holistically drive towards well-being. Capabilities are constitutive elements of well-being and capture the valuable doings and beings that individuals can achieve or become (e.g., being adequately nourished, and being sheltered, able to take a decision, participate in an enjoyable activity, engaged in meaningful livelihood, etc.). The capability approach propagated by Sen (1997) and Nussbaum (2011) in the field of development economic policy brought a human-oriented approach in the development. There are evidences that a wide range of capabilities exhibit statistically significant relations to well-being (Anand et al., 2005). Thus, capability enhances resiliency among the survivors of disaster specifically among the marginalized individuals and groups. Resiliency is understood in individual and community perspective. The individual perspective of resiliency explains two important dimensions; exposure to the adversity and ability, quality of adoption positively (Masten, 2001). In an adverse situation, the survivor will be able to deal with the same based on his/her capacity and resources available. "The ability to spring back from and successfully adapt to adversity is resiliency" (Henderson, 2012). "Resiliency is the capacity to transform oneself in a positive way after a difficult event" (Annan et al., 2003) and in other wards, resiliency is increased or enhanced ability to cope with difficult situations. A resilient community can cope with disturbances or changes and maintain adaptive behavior. For understanding community perspective of resilience, level of adoption at community and population wellness should be measured. This explains "high and non-disparate levels of mental and behavioral health, role functioning, and quality of life in constituent (i.e., disaster-affected area) populations" (Norris et al., 2008; Potangaroa et al., 2015).

7.6   CBPSS IN MAJOR DISASTERS IN INDIA

Some of the major disasters, where professional mental health experts and I were involved in developing and implementing the CBPSS models and these interventions were effectively evaluated for measuring the specific psychosocial outcome, are described here for examining the process of developing protective mechanisms for the vulnerable survivors. In Orissa, Super Cyclone (1999), there was specific program for widowed women and orphan/semi-orphan children (*Sneha Abhiyan*—Campaign of Love (Chachra, 2004; Sekar, 2005). In Gujarat earthquake (2001), there was a similar program named *Sneha Samudaya*—Community of Love, and another major program was for rehabilitation of the disabled (injured, amputees, and paraplegic patients) survivors of disasters (Bhadra, 2015; Ramappa and Bhadra, 2004). In 2002, communal conflict ripped the state of Gujarat and the intervention was initiated through PSS as the core strategy to build peace and for rehabilitation of the survivors through two major projects, *Aman Samudaya* and Gujarat Harmony projects (Bhadra and Dyer, 2011).

In 2004, a workshop was organized on "Mainstreaming Psychosocial Support in Emergencies" in New Delhi, in active participation of the major National, International organizations, and different departments/ministries of India. In the workshop, the importance of community-driven process and PSS was well recognized. Subsequently, in South Asian Tsunami (December 2004), psychosocial issues of the survivors came in forefront immediately and CBPSS was a key intervention model, significantly used by National Institute of Mental Health & Neuro-Sciences (NIMHANS) supported program of CARE-India, and ARC supported program of Indian

Red Cross Society (IRCS) in Tamil Nadu. The PSS program has been effectively implemented in various other disasters, such as Kashmir earthquake rehabilitation (2005), Mumbai train blast school-based intervention (2006) and Uttarakhand Himalayan tsunami (2013). Further, India developed PSS-MHS (mental health services) Guidelines (GOI, December 2009) and National Policy for Disaster Management (GOI, 2009) to ensure effective use of the CBPSS practice.

The significant features of the CBPSS models practiced in these disasters are divided into three dimensions, generic CBPSS, specific PSS strategies in CBPSS for vulnerable survivors, and specialized PSS for specific needs. The discussion has been done with special reference of integrated community recovery project (ICRP) of IRCS supported by ARC. The program was designed with an aim to strengthen recovery through community and school-based interventions and build resiliency among the tsunami survivors (IFRC, 2008). The outcome was envisioned as developing safe, healthy, and competent individuals, families, and communities through community engagement processes, strengthening internal resources of the community and developing better functional community organizations and institutions, such as schools, community training centers, and so forth. The objectives were designed to empower the tsunami-affected communities and schools with better functioning capacity and strengthen local resources and support systems (Hansen, 2008, p. 47; Bhadra, 2013).

7.6.1   GENERIC CBPSS STRATEGIES AND ACTIVITIES

The following are generic activities to introduce PSS in the target communities.

1. Capacity building of the staffs and volunteers on PSS
2. Use of community-level staffs and volunteers for PSS
3. Hand-holding support at the community level by the mental health experts
4. Engaging community as key stakeholders
5. Cultural sensitivity and contextual information
6. Community mobilization
7. Working through the existing support system of health care, child care, and social welfare

8. Promotion of behavior change communication (BCC) through effective IEC materials
9. Use of community folk media

Capacity building considering cascading model was a crucial step to reach out to the survivors at the community by engaging the local volunteers and staffs in the target areas of intervention. Training of the trainers' (TOT) module was developed with effective standardization process to ensure culturally sensitive contextual information with adequate pictorials and local examples. The integrated program of IRCS, supported by ARC developed three modules from master trainer to volunteer's leaders to community facilitators. The training modules precisely covered the psychosocial support, health, and disaster preparedness activities that were effectively practiced at the individual, group, and community levels by engaging the community members as an effective stakeholder in the process. These modules ensured community participation through different mobilization activities, such as participatory community mapping, organizing community events, facilitating group activities, and so on. Engaging the community as stakeholder is a continuum that starts with higher engagement of the external agencies (i.e., IRCS in this case) at the initial stage with a gradual decrease toward the end of the intervention, simultaneously with the gradual increase of community engagements with higher degree towards the end of intervention (Brieger, 2006). Finally, the community control and own the process of recovery. The process of such community engagement in the CBPSS model is shown below (Fig. 7.1).

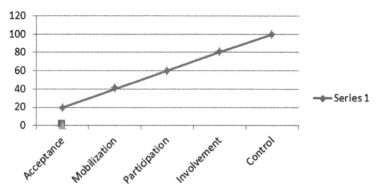

FIGURE 7.1   Levels of involvement
Source: Adapted from The Johns Hopkins University (Brieger, 2006).

While an external agency starts working with a disaster-affected community, the first stage is about being accepted by the local community. As the community accepts the external organization the people start sharing their details and provide support being engaged as volunteer, give space in the community, join in the activities organized, and share opinion. Thus, the community participation becomes evident and in gradual progression, the involvement of the community leaders in the decision-making process happens as desired. The leaders from different community groups such as women, youth, religious leaders, informal leaders (e.g., teachers and doctors), and formal leaders (e.g., political leaders) join in the process as representative of the larger community. The stage of involvement of the local community is vital to push the self-dependency and control by the community themselves. So, at the end, the community control by the community people mark the success of CBPSS model.

With the pace of time, the engagement of external agencies decreased and the community's ownership increased, thereby sustainability of the intervention was encouraged. In this process reaching out to the mass (i.e., the survivor population in the target area) and encouraging them to become an active participant of the integrated program was done through multiple information sharing methods, such as print, nonprint media, folk media activities, by developing community resiliency projects. In this process, continuous support to the staffs and volunteers was provided through regular meetings, refreshment course, and supportive supervision by the mental health experts.

The ICRP of IRCS was uniquely planned to use PSS intervention to build the base for further specific health, psychosocial, disaster prepared-ness programs that were designed to ensure sustainability of the program in long-term through capacity building of the local stakeholders in the insti-tutions and communities (Diaz et al., 2007). The community committee was developed in the participation of local leaders, volunteers to ensure community ownership of the program for conducting various activities, such as campaign, plantation activities, organizing training centers for women, and community park for the children, and so forth. Inclusion of the vulnerable groups was emphasized all through. The model (Fig. 7.2) below explained the process of the same.

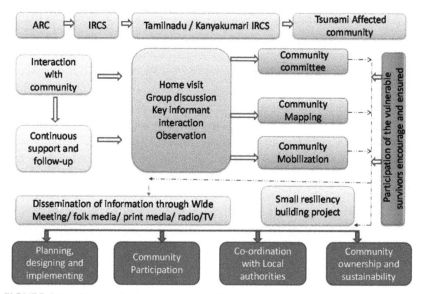

FIGURE 7.2 Community-based psychosocial support model in Integrated Community Recovery Project (ICRP).

Source: (Bhadra and Pratheepa, 2009).

7.6.2 SPECIFIC PSS STRATEGIES IN CBPSS FOR VULNERABLE SURVIVORS

The following activities may be used to integrate vulnerable community members into community activities.

1. Specific training module on PSS for the vulnerable population
2. Working with vulnerable population through individual and family engagement
3. Group interventions with vulnerable groups
4. Enhancing participation of the vulnerable population
5. Possibility of integrated support system for the vulnerable survivors

The factors that contribute towards the vulnerability and marginalization are being tackled through some specific strategies within the CBPSS model and implemented as core features of every disasters intervention. Identification of vulnerable survivors, recognizing the specific nature of problems, and important measures to support the same through program activities is vital for developing protective nets for the vulnerable survivors. Thus, every capacity-building module has a special focus on the vulnerable population and became an integrated practice within the CBPSS model. Training of the trainers' manual of NIMHANS has special focus on vulnerable women, children, aged, disabled with designed session to deal with victims of sexual assault (Lakshminarayana et al., 2002), families with death, persons with severe injuries leading to permanent disabilities, and orphan children (Sekar, et al., 2005).

In the intervention of peacebuilding following the communal conflict in Gujarat the women who were sexually assaulted were identified and contacted by the specially trained women community volunteers. Thus, a circle of support and comfort was rebuilt to boost their confidence and to deal with their feeling of shame and trauma. The intervention was done by maintaining a very high level of confidentiality under the close supervision of experts and found to have a major change in the social and psychological status of the women. Similarly, the young and adolescent boys were supported through intervention in the relief camps, subsequently in their living areas and in the schools through multiple group sessions and life skills education. The children and adolescent survivors could develop a positive outlook about their future and become committed to work towards peacebuilding (Bhadra, 2012).

A model of PSS has worked out rehabilitation of the injured and disabled survivors of Gujarat earthquake. Due to the falling of heavy objects, the number of spinal cord injury leading to paraplegia and amputation of limbs were quite a havoc and caused a severe disabling impact on the disabled persons and their family members, too. Disability caused a severe psychological trauma with physical restriction in movement and dependency on caregivers. Reaching to the survivors in the widely scattered areas of desert land of Gujarat was an important challenge, though the team of physiotherapists, nurses, and social workers visited regularly. The main purpose was to encourage them to accept the physical limitations and develop positive coping to be adjusted with disability status, as the paraplegic and amputees had marked depression and hopelessness. Equally, the caregivers and family members were supported to cope with the stress as they often had to take the anger outburst. *Viklang Bandhu* (friend of disabled) was a cadre of community volunteers, were organized at the community for a regular visit to the disabled survivors. They also mobilized other housing and livelihood service. There were a number of medical supplies, and a temporary

residential care center (*Navajeevan Kendra*) was developed for the physical and psychological rehabilitation of the paraplegic patients (Ramappa
and Bhadra, 2004; Bhadra, 2008, 2015).

The IRCS-ARC modules of tsunami interventions specifically focused
on the underserved families, children of the government schools, and
women. For reaching out to the vulnerable population working with individuals through repeated sessions of psychological first aid for ventilation, enhancing positive coping, and rebuilding individual resources were
vital. With repeated home visits by the trained community volunteers,
group interventions for vulnerable survivors were conducted to develop
their family strength, internal cooperation, and support among the group
members, and to encourage participation in the community activities for
mainstreaming. For example, in the ICRP of IRCS the young children
were considered as one of the vulnerable groups due to a higher prevalence of psychological trauma and mosquito-borne diseases in the coastal
region of Kanyakumari district of Tamil Nadu.

While, protecting this group from traumatic experiences was ensured
through various group activities in the preschool, school interventions, the
protective mechanism for health was developed through community-based
health and first aid (CBHFA) activities, such as health risk identification,
health activity planning, awareness generation, and distribution of insecti-
cide-treated net (ITN) (IRCS-Kanyakumari District Branch, 2008; Singh
and Mini, 2009). The school teachers were trained to conduct various
group sessions with the children and adolescent students to develop positive relationships, making good friends, supporting each other, improving
study habits, working together to make the classroom and school campus
attractive and to showcase their success to the school community at
large. There were special supportive teaching sessions for the students
who missed their regular classes for long, due to the tsunami and living
in relief camps for months together (Bhadra, 2015). In this process the
teachers, parents, students, and community leaders were actively engaged
in designing the specific school activities, such as developing a library,
computer center, making provisions of safe drinking water, hygienic
kitchen, sanitation facilities, and so forth. In evaluation, it was reflected
that the teachers, parents, and other school committee members felt that
they have been able to take active part in the recovery process, and could
mobilize more support from their own community because of the school
intervention program (IFRC, 2008; The Hindu, 2009).

7.6.3   SPECIALIZED PSS FOR SPECIFIC NEEDS OF SURVIVORS

The following are specialized activities that may be used to address for survivors with specific needs:

1. Integration of psychological first aid into the Primary Health Care (PHC) system
2. Mental health care by psychologists or psychiatric social workers and referral
3. Working in schools, informal and religious educational institution for school health recovery

Ingratiating mental health in the PHS service is an important consideration specifically for the mental health service to the common people who often fail to assess services due to social stigma, lack of awareness and continue to suffer from various underlying mental health problems that have very high potential to cause disabling impact in long-term. Primary health care system can treat a maximum number of the mild to moderate case of mental health problems and effectively play a role in facilitating referral to tertiary mental health care. The district mental health program is an important initiative in this direction and this program was extended by the Government of India in the tsunami-affected districts of Tamil Nadu (WHO, 2005).

Any disaster typically has severe disability impact on the children that lead to long-term depression, anxiety, difficulty in concentration, lack of hope, and decreased scholastic perforce (Sheth et al., 2006; Channaveerachari, et al., 2015; Kar, 2009). Thus, school children are a major vulnerable group that every disaster intervention has made a special designed intervention. In ICRP of IRCS the school intervention was designed as per the guidelines of IASC-MHPSS (2007) and INEE (2005). The core features of the program were resiliency building, working through creative expressive activities with the students, developing child-friendly spaces to ensure a sense of place with the school and community, and finally developing a safe-school practice for encouraging disaster preparedness based on the various risks and vulnerability factors. The process of the school of intervention was an assessment of the situation, capacity building of the teachers, developing a proposal for the recovery with the school committee, and implantation through mobilization of from the school community. The teachers were the active implementer of the school intervention with the trained staffs and community volunteers. In the process to develop maximum participation the school committee was developed in each target school with representations of the community leaders, parents, teachers, and few students. The school committee was a body to decide the desired change that would be required for better psychosocial and health development for the students. A model (Fig. 7.3) of this practice is given in the following diagram.

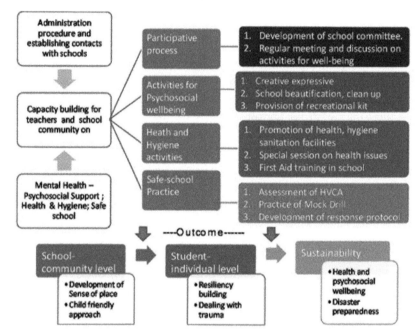

FIGURE 7.3   ARC-supported model of school intervention in ICRP of IRCS.
Source: (Bhadra and Pratheepa 2009).

Similar interventions in the school was conducted in Gujarat riots rehabilitation (Bhadra, 2012), in Kashmir earthquake intervention as well as in the intervention with the children in the schools along the railway tracks in Mumbai train blast (2006). American Red Cross supported Maharashtra state branch of IRCS to conduct a program with the school children with an objective to "facilitate ventilation and ensure regaining confidence among the children in the schools." The children expressed their thoughts about the train blast and about the severe flood that also severely affected their life and psyche in 2006. The intervention was designed for train blast but also responded to the psychological issues due to the severe flood. Following extensive capacity building through participatory and experiential learning methodologies, the teachers conducted various sessions with the students over 6 months and found the school students are expressive to talk about their feelings and could express about their vision, positive spirit, and future development (Tom et al., 2009).

7.7 CONCLUSION

The approach of CBPSS model practiced in the different context of disasters in India marked some key achievements and the existing challenges that still require further exploration for broader humanitarian interest. Though mainly, the chapter explored the issue of protection in the context of human rights for the survivors of natural disasters, the matter is quite deep-rooted in the context of forced displacement, refugee situation in human-made disaster and complex emergencies. In such complicated social situation with a very high threat to the well-being, mental health becomes the most neglected area of relief operations as the need is often not reported by the survivors, and less explored by the humanitarian workers. The mental health complications are proportionally quite high in every disaster situation (Kessler and Ustun, 2011; Dyer and Bhadra, 2013).

It is evident that the intensity of the disaster of both natural and human made, will increase in the coming years and the humanitarian crisis will be widened further with more complicated issues of mental health and well-being. In such situation, the established model of CBPSS can effectively be used to deal with the issues of protection for the vulnerable survivors who often fail to navigate towards the resources, support and remain in the darkness being marginalized. The CBPSS model through promoting individual initiatives, mobilizing groups, family unity, and community empowerment strategies, is an effective, established tool for inclusions of the vulnerable sections of the survivors with protective measures that uphold human dignity. Figure 7.4 explains why PSS is making an impact across the culture and become an established model of practice for survivors of disasters.

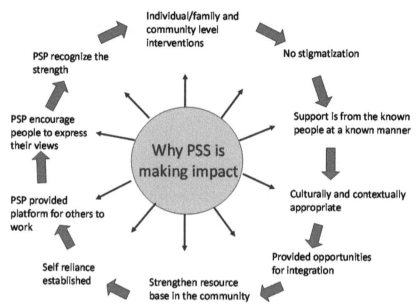

FIGURE 7.4   Multidimensional strength of psychosocial support.

Psychosocial support as a strategy contributes multidimensionally as it ensures interventions at all three levels (individual/family/community), becomes most acceptable by the community to deal with stress and traumatic experiences without any stigma attached to mental health issues. Through PSS activities support is provided by the community volunteers in a culturally sensitive manner, thus, strengthen the resources in the community and establish self-reliance of the survivors after a disaster. Psychosocial support is a strategy that recognizes the strength of the community people, often as a platform facilitates grounds for other interventions.

The CBPSS strategies and process showed distinctively the multifold benefits that could lead the change in the community situation by rebuilding hope, establishing human dignity, crucially through revitalizing the sense of community, belonging, and self-initiatives among the survivors. Eventually, integrated approach has better potential to rebuild the support systems for the vulnerable survivors to facilitate normalization as the approach holistically build an umbrella of support by bridging the gaps between the different sector-specific needs (e.g., livelihood, housing, health, nutrition, sanitation, psychosocial, etc.) and coherence is actualized between the support system and requirements. There are multiple challenges in the way forward to work consistently on CBPSS model due to lack of mental health experts, trained staffs, mental health needs often less recognized, the survivor community is often being treated as a receiver than being engaged as a stakeholder, and finally, the marginalized are served than being empowered. Significantly, CBPSS is now established as an evidence-based model to be widely used in the disaster interventions to protect the survivors from all inhuman sufferings and to reaffirm the commitment towards human rights.

REFERENCES

Aarts, P. G. *Guidelines for Programmes: Psychosocial and Mental Health Care Assistance in (Post) Disaster and Conflict Areas;* International Centre Netherlands Institute for Care and Welfare: Utrecht, The Netherlands, 2000.

Anand, P.; Hunter, G.; Smith, R. Capabilities and Well-Being: Evidence Based on the Sen-Nussbaum Approach to Welfare. *Soc. Indic. Res.* **2005,** *74*(1), 9–55.

Annan, J.; Castelli, L.; Devreux, A.; Locatelli, E. *Training Manual for Teachers, AVSI;* Edinburgh: AVSI. 2003 йил 9-January from http://www.forcedmigration.org/psychosocial/papers/WiderPapers/Teachers%20manual.pdf. (accessed 2005).

Barber, R. Localising the Humanitarian Toolkit: Lessons from Recent Philippines Disasters. In *Natural Disaster Management in the Asia-Pacific: Policy and Governance;* Brassard, C., Giles, D. W., Howitt, A. M., Eds.; Springer Japan; Tokyo, 2015; pp 17–32.

Barker, R. L. *The Social Work Dictionary;* 5th ed.; National Association of Social Worker: Washington, DC, 2003.

Bhadra, S. Gujarat Earthquake Experience. In *JTCDM Working Paper;* Jaswal, S., Joseph, J., Eds.; Jamsetji Tata Centre for Disaster Management: Mumbai, Maharastra, India, 2008; p 35.

Bhadra, S. Psychosocial Support for the Children Affected by Communal Violence in Gujarat, India. *Int. J. Appl. Psychoanalytic Stud.* **2012,** *9*(3), 212–232.

Bhadra, S. Community Based Psychosocial Support Programme for Resiliency Building in Tsunami Rehabilitation of Kanyakumari District. *J. Soc. work, Special Issue on Building Resilient Communities: Communitarian Social Work,* **2013,** *3*(8), 66–86.

Bhadra, S. Failure and Hope: Living or Not with Disability after the 2001 Gujarat Earthquake. In *Disability and Disaster;* Kelman, I., Stough, L. M., Eds.; Palgrave Macmillan: New York, 2015a; pp 61–68.

Bhadra, S. Human Trafficking in Humanitarian Crisis of Natural and Manmade Disasters in India. *Soc. Work J. Assam Cent. Univ.* **2015b,** *3*(2), 44–56.

Bhadra, S. Mental Health and Psychosocial Support for Tsunami Affected School Children in Kanyakuumari District, South India. In *Social Work Practice in Mental Health:*

Cross-Cultural Perspectives; Francis, A. Ed.; Allied Publishers Pvt Ltd.: New Delhi, pp 105–120; 2015c.

Bhadra, S. The Marginalized Survivors in the Process of Disaster. In *Multidisciplinary Handbook of Social Exclusion and Human Rights;* Chakraborty, S., Nanjurda, Eds.; Aaya Publication: New Delhi, 2016; pp 243–263.

Bhadra, S.; Pratheepa, C. M. *Strengthening Communities and recovery Through Psychosocial Support;* 2009. National Institute of Disaster Managment: 2nd India Disaster Managment Congress: http://nidm.gov.in/idmc2/PDF/Presentations/Psycho_Social/ Pres3.pdf. (accessed June 26, 2013).

Bhadra, S.; Dyer, A. R. Psychosocial Support for Harmony and Peace Building: Rebuilding Community in Gujarat. In *Peace from Disasters—Indigenous Initiatives Across Communities, Countries and Continents;* Peacebuilding, H. U., Ed.; Hiroshima University: Hiroshima, 2011; pp 97–104. http://home.hiroshima-ac.jp/hipec/.

Brieger, W. R. *Johns Hopkins Bloomberg School of Public Health;* 2006. http://ocw.jhsph. edu/courses/socialbehavioralfoundations/PDFs/Lecture15.pdf. (accessed Jan 4, 2014).

Chachra, S. Disaster and Mental Health in India: An Institutional Response: Action Aid India. In *Disaster Mental Health in India;* Diaz, J., Murthy, R. S., Lakshminarayana, R., Eds.; Indian Red Cross Society: New Delhi, 2004; pp 151–160.

Channavecrachari, N. K.; Raj, A.; Joshi, S.; Paramita, P.; Somanathan, R.; Chandran, D.; Badamath, S. Psychiatric and Medical Disorders in the after Math of the Uttarakhand Disaster: Assessment, Approach, and Future Challenges. *Indian J. Psychol. Med.* **2015,** *37*(2), 138–143.

Companion, M. *Disasters' Impact on Livelihood and Cultural Survival: Losses, Opportunities, and Mitigation;* CRC press-Taylor and Francis Group: Boca Raton, 2015.

Dhar, N.; Chaturvedi, S.; Nandan, D. Spiritual Health Scale 2011: Defining and Measuring 4th Dimension of Health. *Indian J. Community Med.* **2011,** *36*(4), 275–282. DOI: 10.4103/0970-0218.91329.

Diaz, J. O.; Bhadra, S.; Krishnan, P. Psychosocial Support as a Platform for an Integrated Development Program. *Coping with Crisis,* **2007,** *2*, 4–7.

Dyer, A. R.; Bhadra, S. Global Disaster, War Conflict and Complex Emergencies: Caring for Special Population. In *In 21st Century Global Mental Health;* Sorel, E., Ed.; Jones and Bartett Learning: Washington; pp 171–209, 2013.

GOI. *National Disaster Management Guidelines: Psychosocial Support and Mental Health Services in Disasters;* National Disaster Managment Authority, Government of India: New Delhi, 2009.

Hansen, P. *Psychosocial Interventions A handbook;* Agen, W., Ed.; International Federation Reference Centre, for Psychosocial Support: Copenhagen, 2008.

Henderson, N. *The Resiliency Workbook:Bounce Back Stronger, Smarter and with Real Self-Esteem;* Resiliency in Action: California, 2012.

IASC. *Protecting Persons Affected by Natural Disasters: Operational Guidelines on Human Rights and Natural Disaster;* Inter-Agency Standing Committee: Washington, DC, 2006a.

IASC. *Women, Girls, Boys and Men: Different Needs—Equal Opportunities;* Inter Agency Standing Committee, United Nation: Geneva, 2006b.

IASC. *Guidelines on Mental Health and Psychosocial Support in Emergency Settings (MHPSS);* IASC: Geneva, 2007.

IASC. *Humanitarian Action and Older Persons: An Essential Brief for Humanitarian Actors;* IASC: Geneva, 2008. http://www.who.int/hac/network/interagency/iasc_advocacy_paper_older_people_en.pdf (accessed May 23, 2016).

ICRC. *Resolutions of the Geneva International Conference. Geneva,* 2008. October 26–29, 1863. Treaties, States Parties and Commentaries.   https://www.icrc.org/ihl/INTRO/115?OpenDocument (accessed May 2, 2016).

IFRC. *Psychosocial Framework 2005–2007;* International Federation of Red Cross (IFRC): Geneva, 2005.

IFRC. *Federation-wide Tsunami Semi-annual Report 2004–2008: India Appeal No. 28/2004.* 2008, July 21. http://reliefweb.int/report/india/federation-wide-tsunami-semi-annual-report-2004-2008-india-appeal-no-282004.

Indian Red Cross Society (IRCS), Kanyakumari District Branch. *Long Lasting Insecticide Net (LLIN) Distribution Final Report.* Nagercoil: IRCS-Kanyakumari District Branch. 2008a. http://www.againstmalaria.com/images/00/05/5536.pdf (accessed May 13, 2013).

Indian Red Cross Society (IRCS), Kanyakumari District Branch. *Teacher's Manual; Integrated Recovery Programme in School for Children;* Indian Red Cross Society (Supported by American Red Cross): Nagercoil, 2008b. http://www.pscentre.org/wp-content/uploads/gallery/2013/03/TeachermanuaKKTRP.pdf.

Kar, N. Psychological Impact of Disasters on Children: Review of Assessment and Interventions. *World J. Pediatr.* **2009,** *5*(1), 5–11.

Kelman, I.; Stough, L. M. Disability and Disaster. In *Disability and Disaster;* Kelman, I., Stough, L. M., Eds.; Palgrave Macmillan: New York, 2015; pp 3–14.

Kessler, R.; Ustun, T. *The WHO World Mental Health Surveys: Global Perspectives on the Epidemiology of Mental Disorders;* Cambridge University Press: New York, 2011.

Lakshminarayana, R.; Sen Dave, A.; Shukla, S.; Sekar, K.; Murthy, R. Psychosocial Care by Community Level Workers for Women. *Information Manual—4;* Books for Change: Bangalore, Karnataka, India, 2002.

Larson, J. S. The World Health Organization's Definition of Health: Social versus Spiritual Health. *Soc. Indic. Res.* **1998,** *38*(2), 181–192.

Leidig, M.; Teeuw, R. M. Quantifying and Mapping Global Data Poverty. *PLoS One,* **2015,** *10*(11), 1–15. DOI:10.1371/journal.pone.0142076.

Masten, A. S. Ordinary magic: Resilience Processes in Development. *Am. Psychol.* **2001,** *56*(3), 227–238.

Norris, F.; Stevens, S.; Pfefferbaum, B.; Wyche, F.; Pfefferbaum, R. Community Resilience as a Metaphor, Theory, Set of Capacities and Strategy for Disaster Readiness. *Am. J. Community Psychol.* **2008,** *41*(1–2), 127–150.

Nussbaum, M. *Creating Capabilities: The Human Development Approach;* Harvard University Press: Cambridge, MA, 2011.

Potangaroa, R.; Santosa, H.; Wilkinson, S. Disaster Management: Enabling Resilience. In *Lecture Notes in Social Networks: Disaster Management—Enabling Resilience;* Masys, A., Ed.; Springer: New York, 2015; pp 227–266.

Ramappa, G.; Bhadra, S. Institutional Responses—Oxfam (India): Psycho social Support Programme for Survivors of the Earthquake. In *Disaster Mental Health in India;* Lakshminarayana, R., Murthy, R. S., Diaz, J. O., Eds.; Indian Red Cross Society: New Delhi, 2004; pp 140–160.

Sekar, K. Orissa Supercyclone. In *Disaster Mental Health in India;* Laskshminarayana, R., Murthy, S. R., Diaz, J. O., Eds.; Indian Red Cross Society: New Delhi, pp 110–118; 2005.

Sekar, K. Human Rights and Disaster: Psychosocial Support and Mental Health Services. In *Mental Health Care and Human Rights;* Nagaraja, D., Murthy, P., Eds.; NHRC and NIMHANS: Bangalore, pp 243–266, 2008.

Sekar, K.; Bhadra, S.; Jayakumar, C.; Aravindraj, E.; Henry, G.; Kumar, K. K. *Facilitation Manual for Trainers of Trainees in Natural Disaster;* NIMHANS and Care India: Bangalore, 2005.

Sen, A. K. Distinguished Guest Lecture: From Income Inequality to Economic Inequality. *South. Econ. J.* **1997,** *64*(2), 384–402.

Sheth, A.; Sanyal, S.; Jaiswal, A.; Gandhi, P. Effects of the December 2004 Indian Ocean Tsunami on the Indian Mainland. *Earthquake Spectra,* **2006,** *22*(S3), S435–S473.

Singh, N.; Mini, J. *Important of Community Health Assessment Through Community Based Approach in Post Disaster Period.* National Institute of Disaster Managment, 2nd India Disaster Management Congress. Nov 8–9, 2009. http://nidm.gov.in/idmc2/PDF/Presentations/PHE/Pres3.pdf (accessed June 26, 2013).

The Hindu. *Red Cross Comes to the Aid of Affected People.* The Hindu. Sept 13, 2009. http://www.thehindu.com/2009/09/13/stories/2009091350990300.htm (accessed Jan 23, 2013).

The Inter-Agency Network for Education in Emergencies (INEE). *Minimum Standards for Education in Emergencies, Chronic Crises and Early;* INEE Network: London, 2005.

The SPHERE project. *Humanitarian Charter and and Minimum Standards in Humanitarian response;* Oxfam Publishing: Oxford, UK, 2004.

The SPHERE project. *Humanitarian charter and Minimum Standards in Humanitarian Response;* The SPHERE Project: UK, 2011.

Twigg, J. (2004). Disaster risk reduction: Mitigation and preparedness in development and emergency programming. London, UK: Humanitarian Practice Network (HPN). http://www.ifrc.org/PageFiles/95743/B.a.05.%20Disaster%20risk%20reduction_%20 Good%20Practice%20Review_HPN.pdf

Tom, J. J.; Bhadra, S.; Modi, H. N. Psychosocial Response in Mumbai Train Blast through. *National Disaster Managment Congress-II* (p. 2). New Delhi: National Institite of Disaster Management, 2009. http://nidm.gov.in/idmc2/PDF/Abstracts/PSC.pdf (accessed May 12, 2016).

UNICEF. *Child Trafficking,* 2012. http://www.unicef.org.au/About-Us/What-We-Do/ Protection/Child-Trafficking.aspx (accessed March 29, 2015).

UNISDR. *Fourth Asian Ministerial Conference on Disaster Risk Reduction.* Incheon, Korea: UNISDR, 2010. http://www.unisdr.org/files/16172_finalincheondeclaration-draftingcom1.pdf. (accessed May 30, 2016).

United for Human Rights. *History of Human Rights*, 2016. United for Human Rights. http://www.humanrights.com/what-are-human-rights/brief-history/declaration-of-human-rights.html (accessed May 2, 2016).

United Nations. *Convention of the Rights of Persons with Disabilities.* United Nations: Geneva, 2007.

UNOHCHR. *The Core International Human Rights Instruments and their monitoring bodies,* 2016. United Nations Human Rights: http://www.ohchr.org/EN/ProfessionalInterest/Pages/CoreInstruments.aspx (accessed May 30, 2016).

Walker, B.; Westley, F. Perspectives on Resilience to Disasters Across Sectors and Cultures. *Ecol. Soc.* **2011,** *16*(2), 1–4. file:///C:/Users/hp/Downloads/ES-2011-4070.pdf.

WHO. *Strengthening mental health promotion (Fact sheet, No. 220).* World Health Organization: Genva, 2001.

WHO. *Mental Health and Psychosocial Relief Efforts after the Tsunami in South-East Asia;* World Health Organization: New Delhi, 2005. http://www.searo.who.int/entity/emergencies/documents/mental_health_and_psychosocial_relief_efforts_fulltext.pdf?ua=1 ((accessed Jan 13, 2014)

Zastrow, C. H. *Evaluating Social Work Practice;* Cengage Learning, India Edition: New Delhi, 2010.

CHAPTER 8

IMPLEMENTATION OF COMMUNITY-BASED PSYCHOSOCIAL SUPPORT IN SOUTH ASIA

SATYABRATA DASH

Senior Psychiatrist, Apollo Hospitals, Bhubaneswar, Odisha, India

CONTENTS

8.1 INTRODUCTION

India is the land of disasters and multiple responses (Gupta, 2006). The first Disaster Mental Health programs in India began after the 1981 Bangalore Circus Fire tragedy. National Institute of Mental Health and Neurosciences (NIMHANS) took the lead and provided psychosocial support to the survivors and bereaved families. Subsequent to that NIMHANS has been a pioneer in psychosocial support post disasters in India. From the Bangalore circus fire (1981), Bhopal gas tragedy (1984), Bombay riots (1992–1993), Marathwada earthquake (1994), Odisha super cyclone (1999), Gujarat earthquake (2001) to Gujarat riots (2002) NIMHANS had been the lead agency in India. The American Red Cross (ARC) had its own national disaster mental health programs in the United States since 1972 with the formation of the mental health emergency section in National Institute of Mental Health (NIMH) but the international services of the ARC did not have a developed disaster mental health response program. The ARC international services started disaster mental health programs in the late 1990s and early 2000s with programs in Nicaragua, Gujarat, and El Salvador. In India, post the Gujarat Earthquake in 2001 the ARC and the Danish Red Cross conducted initial assessments and agreed that there was a need for trauma counseling as well as some psychosocial activities for children and the elderly to alleviate fear (Dodge, 2001). The ARC international services developed a program mirroring the disaster mental health program of the domestic services (Jacobs, 2001). As a result of the Gujarat riots, the ARC brought in a consultant to address needs of the survivors. The program was psychosocial in nature and used community mobilization as its theoretical base (Mathews, 2002). The program matched existing ARC/Spanish Red Cross programs in Central America and was initiated after hurricane Mitch (Prewitt Diaz and Biel, 2002).

In the meantime, there was another change that was happening in regard to mental health in post-disaster context. The concept of disaster mental health with mental health professionals and volunteers were giving way to community-based psychosocial support programs (PSPs) within the Red Cross Red Crescent (RCRC) responses, as exhibited by the Turkey earthquakes in 1999 and the Afyon earthquake February 2002 that killed over 31,000 people. Also in India, community-based PSPs were held following the Odisha super cyclone of 1999 by Action Aid International.

By 2002, the Indian Red Cross, Spanish Red Cross, and the ARC identified the need for continued psychosocial support to alleviate fear in the coastal region in Odisha (Prewitt Diaz et al., 2003). The program focused on two basic conduits: (1) community mobilization in communities and (2) using the children to develop preparedness activities in schools (Prewitt Diaz et al., 2004a). The idea was to develop a model in the context of the region that could be applied to disasters in countries in Asia. Some years later, the learning from these programs was transferred to assist the tsunami-affected people in Maldives and Bangladesh.

8.2   LESSONS LEARNED FROM THE CBPSS DEVELOPMENT IN ODISHA AND USE IN LATER DISASTERS

The PSP (named "Disaster Mental Health Psychosocial Care"—DMHPC), in Odisha, was developed long after the Odisha super cyclone of 1999 during the reconstruction and rehabilitation phases of the disaster in 2003. The objectives comprised: (1) developing capacity of villages, (2) enhancing the capacity of schools to develop crisis response plans, (3) developing the capacity of community volunteers to offer psychological first aid (PFA), and (4) enhancing the capacity of Indian Red Cross State Branch to develop manage DMHPC programs by developing skilled volunteers (Dash et al., 2003). Some super cyclone affected villages and schools were selected after assessments by incorporating community visits, focused groups, and interviews. The program personnel showed an enthusiastic desire to participate in the program activities, monitor their growth, and report on the accomplished achievements. The comprehensive program activities were implemented in 30 villages with a total population of 13,352 from the coastal area of Puri district, self-selected for the community project and activities (Dash, 2003). A total of 180 schools from four districts including the district chosen for the community project were taken up for the school project and activities (Dash, 2004).

In the target villages, 148 volunteers, known as "community facilitators (CFs)," were trained to provide PFA to the disaster-affected populations, carry out community resilience-building activities, and draw community crisis response plans through participatory methods (See program monthly reports and final external evaluation by the Indian Red Cross, 2004). There were 16 paid staff people that developed and administered the program over a 2-year period.

The crisis intervention specialist (CIS) training component for enhancing the capacity in Odisha for psychosocial support programming trained 66 participants (all participants had a bachelor's degree or higher) to oversee community activities by the trained CFs and school teachers. Most of the participants were either affiliated with the *Family Counseling Centers of the State Social Welfare Advisory Board covering 15 districts of Odisha or the Indian Red Cross Disaster Preparedness and Management Program (IRCS-ODMP), and Spanish Red Cross-Assisted Program.*

Additionally, PFA (a two-day basic training) was provided to 900 community level volunteers in 16 districts of Odisha. Information, education, and communication materials and training materials were developed that were linguistically, culturally, and contextually appropriate to the target population. All these trained resources and materials, program development and design were handed over to the state branch of the Indian Red Cross for sustainability when the program was completed in 2004 (Fig. 8.1).

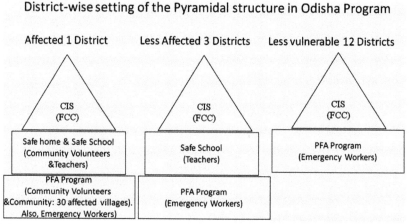

District-wise setting of the Pyramidal structure in Odisha Program

FIGURE 8.1   District-wise setting of the pyramidal structure in Odisha program, (Dash, 2003).

There were several factors that influenced the success of the program. The key elements to the PSP were (1) use of the communities in planning, implementing, and monitoring the progress of their own programs, and (2) locally developed information education and communication and training materials. These two strategies made the program well acceptable because it used local understanding and acceptability in implementation (See program monthly reports and final external evaluation by the Indian Red Cross, 2004). The number of volunteers and costs incurred made the implementation financially feasible and sustainable to carry out in large populations. The materials left behind were helpful in continuing the program after the external funding ended (IRCS, 2010). Another important strength of the PSP was the use of the local organizations, both government and nongovernment as a conduit to enhance the grassroots skills of people. The skills of the already existing human resources put in place by the local government for its communities (alone or with the help of

United Nations (UN) agencies) were enhanced for disaster mental health strengthening and disaster response system of the government (IRCS-ODMP-2, 2012). This element was helpful in being readily acceptable by the local authorities as it was in line with their programs and lead to quicker implementation due to the convenient selection process of trainees and quicker paperwork. The other element was the two-pronged approach to the target population—children through schools and general population through community volunteers Prewitt Diaz, Murthy, Lakshminarayana (2004). At the end of the intervention, the target communities and schools had received training, materials, and had developed local response plans in event of any contingency.

The Odisha program was of very short duration, about 18 months, and was experimental in nature being the earliest of its kind. The program's technical design, targets, liaison, monitoring and evaluation, and human resource allocation were extremely challenging. The school component could never really be implemented as planned. The lessons learnt in the program were: (1) it is not possible to make desired impact on the target population especially in resilience building within program period of less than 3 years, (2) the activities in the schools and the communities needed to be technically predesigned with scope of local modification to produce effect rather than being spontaneously designed by the target population, (3) the program needed the test of efficacy in emergency situations. Although, during the program period in India, there were two special occasions where the ARC technical team had to intervene in an emergency situation—fire tragedy at Kumbakonam in Tamil Nadu and lightning tragedy in a school at Chaupali, Odisha.

8.3   CASE STUDY 1: PSYCHOSOCIAL SUPPORT IN THE REPUBLIC OF MALDIVES

8.3.1   THE FIRST FOUR WEEKS

In December 2004, while the India Red Cross Delegation was closing down, South Asia was jolted by the 26th December Banda Aceh Earthquake and the tsunami that hit seven Asian countries killing approximately 230,000 people and affecting millions. One of the seven Asian countries that were hit by the tsunami was the Republic of Maldives. The Maldives is an archipelago of islands of which 186 are inhabited and distributed in 26 atolls.

Sixty-seven of the inhabited islands in seven atolls were flooded, 82 people lost lives and 26 were missing but the destruction and economic loss were massive. The Maldives did not have a Red Crescent Society of its own so when the International RCRC Movement moved to the Maldives to provide emergency disaster response services, it looked for support for psychosocial support providers in the region within the Red Cross and Red Crescent Movement. One such national society partner was the ARC disaster mental health and psychosocial support expert team in New Delhi, India.

The Maldives psychosocial support was an emergency intervention that had to be done. The expert group in conjunction with the Maldivian emergency management agency planned a two-pronged approach. First, counselors would be trained in PFA, and volunteers from the islands would be recruited to form emotional support brigades (ESB). Second, teachers would be trained to provide psychological support to children in their respective schools (Fig. 8.2).

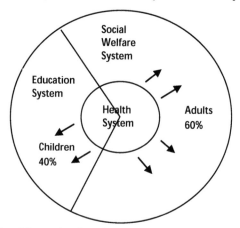

FIGURE 8.2   Details of the mode of psychosocial service delivery.
Source: (Dash, 2006).

Originally, the PSP team was attached to the International Federation of Red Cross and Red Crescent Societies (IFRC)/field assessment coordination teams (FACT). The Maldives government—the National Disaster Management Centre—had already set up a psychosocial response unit—social support and counseling services (SSCS)—and had selected 70 counselors to move to the islands and help the affected population. The first lesson learned in the Odisha DMHPC program was to contextualize technical components to the local reality. The lesson of designing technical components to the locally existing resources and networks worked very well and got an immediate push from the local authorities.

The second lesson learned was that materials had to be culturally and linguistically accepted by the affected people. The designed program modules from India were quickly adapted to the Maldivian language and culture. The third is to train local resources to deliver the materials and programs. A group of counselors was trained and sent to the islands in small groups to provide PFA within the first 2 weeks of the tsunami. Thirteen of the counselors were trained as PFA trainers with 300 volunteer hours committed by each for direct services and training in PFA as part of capacity building. Fifty-seven counselors were trained in PFA for direct services.

The PFA trained counselors visited the affected islands in small groups spending 1–2 days at a time and covering overall 31 islands. In these communities, they provided direct services to the population (adults, children, and elderly) by providing PFA, educated the community, in focused groups, on the reactions being normal in an abnormal situation; the necessity of sharing and going back to daily routines. The counselors developed a cadre of community volunteers on PFA techniques and the mechanics of continuing psychological support activities after the departure of the counselors. The community volunteers (proper Emotional Support Brigades) were listed in 23 islands and were followed up by the counselors through the toll-free helpline. During the recovery period, approximately 22,238 people received psychological support services through PFA program, as reported by the islands to the central government. This number included people that received services directly from PFA trained counselors and indirectly through psychological support activities conducted by ESB built by the counselors.

In the school program, six teachers from Male were included in the training teams for the planning and cultural adaptation of the teachers training materials. These teachers were also included in the facilitation of the training sessions as part of capacity building. Three teams with two Maldivian teachers under the supervision of one ARC trainer conducted trainings in the atolls.

Total 16 training sessions were organized in various atolls with the support of ARC, SSCS, and the United Nations International Children's Emergency Fund (UNICEF). The SSCS organized the trainings, the ARC provided the trainers and materials, and the UNICEF provided the logistics for the trainees to come to the training island and the stay. The 2-day training sessions included 321 teachers from 226 schools in 20 atolls. These trained teachers disseminated information on psychological support to their fellow teachers and parents and provided direct psychological support to the children of their schools. Further, these teachers were targeted to be trained in the long-term psychosocial support program for the schools, the Safe School Program. School chests containing kits for recreational and expressive activities for children were distributed to the schools where the teachers' trainings were conducted with a demonstration to the teachers undergoing training. Through this component, the estimation was done that 89,338 children were estimated to have received psychological support in due time (Maldives Emergency Management Agency, 2006).

8.3.2   THE NEXT THREE AND HALF YEARS

During the recovery and reconstruction phase, a long-term program was initiated in all inhabited islands of seven atolls of Maldives—Raa, Meemu, Dhaalu, Thaa, Laamu, Gaafu Alifu, and Gaafu Dhaalu. A total of 65,155 beneficiaries spread out over 76 island communities and 114 schools were included in the program.The long-term program was designed to develop the capacity at the national level to implement programs that enhance and promote community psychosocial well-being by strengthening community resilience; to enhance psychosocial well-being of school environment, and to disseminate the best practices and lessons learned widely for future psychosocial programming.

The program was designed to build national capacity to implement PSPs by developing a pool of trained volunteers from a variety of groups including professors, teachers, community volunteers, and other professionals. These individuals were trained at different levels depending upon their training needs and role within the program. The different levels are—CF, school facilitator (SF), crisis intervention technician (CIT), and CIS. The CIT and CIS programs had a built-in "how to train" component in them and volunteer hours commitment in addition to a two-day PFA training for emergency responders. The total of 100 CIS, 73 CIT, and

221 psychological first aiders (among the emergency responders) were trained. Linguistically and culturally appropriate materials such as training manuals, posters, leaflets, and educational videos were developed.

In the target communities, the trained resources conducted planned activities. The purpose of these activities had three branches: (1) direct services, (2) information and education activities, and (3) resilience building.

1. *Direct service*: Addressed the psychological needs through a community-based approach including facilitating referrals for survivors requiring clinical help to the health system. Sample activities included sharing and other self-care activities (including recreational, memorial, and spiritual activities) within/between communities directed towards ventilation of emotions and reduction of stress.

2. *Information and education*: Provided information and education on psychosocial support and information sharing activities such as the distribution of trifolds, posters, and leaflets; and exhibitions and educational programs in the media (see below).

3. *Resilience-building activities*: Developed positive social interactions within and between communities in the backdrop of lack of cohesion between the different island communities and big internally displaced population. These activities included social and cultural activities directed towards strengthening of bonds within/ between communities such as celebrations of a culturally important day together by the community/communities; marking homecoming of internally displaced persons (IDPs) into host community and weekend fairs with the participation of various communities. It also included activities directed towards addressing community challenges such as drawing up community crisis response plans and developmental planning activities such as repairing of damaged infrastructure (roads, libraries, playgrounds, and parks) or building new community assets; and other concrete, purposeful, common interest activities that involve enhancing social cohesion within/ between communities.

The program trained one CF for every 50 community members, and these CFs extended psychosocial support to 66,136 people in the 7 target atolls. Overall 1137 CFs were trained, and 819 community resilience activities and 23 resilience-building projects were conducted.

The school component trained 212 teachers in 142 schools and preschools in the 7 target atolls and conducted 299 psychosocial activities for 26,353 children. The program aimed to provide not only a physically secured school environment but also a space for dialogue between teachers and students, and an environment where students would be encouraged to become actively engaged in all school activities. This was achieved by training 1–3 teachers in each school as school PSP facilitators who then conducted activities to foster a positive relationship between teachers, students, and parents.

With the assistance of program staff, they also provided information to teachers to facilitate creative and expressive activities within the curriculum that focused on building resilience in the students. Predesigned classroom activity manuals were developed by the program for three age groups (preschool, age 6–11, and adolescents). The classroom activities were in the form of classroom games with the aim of developing social interaction skills, social cohesion, emotional regulation, empathy, self-esteem and overcoming fears; and the creative expressive activities involved similar purposes through use of mediums such as colors, crayons, and clay. Metallic boxes containing the exact required materials to conduct classroom activities as per the abovementioned manuals for an academic year with the list of materials and manuals called "school chests" were provided to all target schools for two consecutive years with enough materials to conduct classroom activities for one academic year. To ensure sustainability of these classroom activities, meetings were held with the Minister of Education and other senior staff of the ministry to expand the program to all schools in the future. Other activities include marking special days such as Children's Day, and information dissemination sessions for parents and recreational activities.

8.3.3   INFORMATION DISSEMINATION

Based on success stories and information on psychosocial support and well-being 17 videos, 32 booklets and leaflets were developed and distributed in the program islands and ministries. Ministry of Education, National Disaster Management Center, and the PSP Reference Center of IFRC uploaded ARC PSP materials to their websites.

Program staff made several presentations at local and international forums—at government ministries and among mental health professionals and shared information about the program and lessons learned. Articles have been published in the Asian Disaster Preparedness Center (ADPC) newsletter and the program has been presented at ICAP (International Congress of Applied Psychology) in Greece, the World Conference on mental health promotion in Oslo, and the public health in complex emergencies (PHCE) training course conducted by ADPC in Bangkok. Ministry of Education officials also presented the school component of the program at the inter-country workshop on mental health promotion and life skills education in New Delhi in December 2007.

This program was undertaken for 3.5 years as per our previous experience in India. At the end of the program, an evaluation was done for the efficacy of the program. The evaluation indicated—A total of 66,136 beneficiaries spread out across 76 islands in 7 atolls of the Maldives benefited through the program. The program trained 1830 volunteers at various technical levels to provide psychosocial support to their communities.

The program is believed to be of over 3.5-year duration to strengthen social capital in the affected communities as seen in the survey, but it will still require a longer period for communities to practice providing psychosocial support to each other and teachers to provide psychosocial support to school children.

Below are some excerpts from the evaluation report obtained from the community through community participatory methods:

> *"Beneficiaries and government partners indicated that since the tsunami, island communities have gained a new level of awareness about their need for increased psychosocial support. The psychosocial project appears to have contributed to this awareness at the island level and governmental levels through resilience building activities, school activities, and trainings."*

> *"Training of community facilitators appears to have been useful in some cases, in dealing with inter-island conflicts (between IDP and host communities and between political groups)."*

> *"Community facilitators were able to help children deal with anxiety about conflict. Resettled IDP youth identified that resilience activities were helpful."*

Most of the resources such as the technical materials were handed over to the government of the Maldives, and the Maldivian Red Crescent Society that formed in parallel to the disaster response programs of the Red Cross and Red Crescent societies.

8.3.4 SUMMARY

The Maldives program validated the best practices of the Odisha program and helped in the smooth implementation of the program. Additionally, the Maldives emergency phase program gave a trial run to an ideal psychosocial programming scenario with successful implementation of emergency psychosocial response and development of a strong base platform for the development of the long-term program.

In the long-term program, the trainings were reset at different levels, being a country-level program. The trainings were ranging from CISs (for country-level program managers), CIT, CFs, and SFs to the psychological first aiders. As a step further to widen its reach to the population the program trained community volunteer to the general population to a ratio of 1:50.

The materials developed in the Maldives program were more advanced than those in the Odisha program. Detailed activity books for the school teachers to build resilience in children were developed and materials to carry out the same—called school chests—were provided to the target schools.

The need was felt to further fine-tune the community program with detail outline of activities to be conducted in the communities which were not developed in the Maldives program. Also, there felt a need for developing trainer's manual and training package in addition to the manual received by the trainee as part of sustainability. The evaluation report indicated that beneficiaries felt the need to be more involved in the development of the program, and the monitoring and evaluation need to be more advanced with the inclusion of qualitative indicators.

8.4 CASE STUDY 2: THE PROGRAM IN BANGLADESH: MODIFICATIONS AND RESULTS

While the Maldives program was coming to an end by early 2008, a region of Bangladesh was devastated by Cyclone Sidr in November 2007 and the International Federation of the Red Cross and Red Crescent Societies decided to carry out a PSP for the cyclone-affected population in four districts of Bangladesh. The author who was the country manager of the ARC PSP in the Maldives moved to Bangladesh as a delegate of the International Federation to develop the PSP. The Bangladesh program, like the Odisha program, did not have an emergency phase but it started in the immediate next phase of the disaster cycle—the recovery phase.

In Bangladesh, there were multiple humanitarian agencies already on the ground prior to the cyclone, unlike the Maldives; and various components of the psychosocial support were already being covered by other agencies. The training of health care professionals on basic mental health management and referrals was being conducted by the World Health Organization (WHO). The teachers' training on psychosocial support in schools was being addressed by the Save the Children Fund. The IFRC assumed responsibility for the community component of psychosocial support in target communities in four affected districts of Bangladesh covering 20,000 households and the capacity building of Bangladesh Red Crescent Society (BDRCS) and its staff and volunteers in providing psychosocial support to fellow volunteers, fellow staff, and communities. The program lasted a short period of time and closed within a year due to management and liaison issues.

A total of 20,000 households which included 49 communities in four districts—Bagerhat, Barguna, Pirojpur, and Patuakhali were served during the program through direct services. Additionally, in nine districts—Bagerhat, Barguna, Pirojpur, Patuakhali, Barisal, Jalokati, Satkhira, Khulna, and Bhola, volunteers from BDRCS were provided for psychosocial support and their capacity in providing psychosocial support enhanced.

8.4.1   CAPACITY BUILDING

The program had association with the Department of Psychology, Dhaka University with a view to develop enhanced capacity of program designers in the country of operation and 22 master trainers received staff development, who subsequently went out to the districts with the program staff and provided training of trainers (TOT) and to the BDRCS community volunteers as part of their volunteer hours.

The component provided training to the BDRCS personnel and volunteers in community-based psychosocial support which included sessions on PFA, self-care and, community social recovery. 464 BDRCS volunteers from nine districts (including the four districts targeted for community programs) were trained in 16 sessions. This ensured the personnel and volunteers getting information on traumatic stress and psychological reactions, basic nontechnical psychological support intervention (PFA), identification of individuals requiring referral services, phases of the psychosocial response of communities to disasters, and the interventions for community psychosocial recovery. Both the process of the training (3 days of interactive engagement) and learning helped the personnel and volunteers overcome their own experience of traumatic stress. Furthermore, the training enhanced their knowledge and capacity in addressing such issues in the communities in future disasters. The volunteers were followed up by the program team, on request, providing psychosocial support services and linking the volunteers to support systems whenever required.

About 40 of the BDRCS district-level volunteers from the target districts for the community were selected based on their performance for serving as TOTs and received an additional day (4th day) of training on "training methodology" in the same setting. These TOTs built the capacity of the community-level volunteers and also facilitated the psychosocial support activities through the community level volunteers. Training and subsequent experience in PSP implementation in the communities for a year ensured the building of capacity of these volunteers and the BDRCS unit in the district.

The enhancement of the psychosocial well-being of the most vulnerable cyclone Sidr-affected 49 target communities in the 4 districts was done by training community level volunteers on psychosocial support and conduction in their respective communities.

8.4.2 SELECTION AND TRAINING OF VOLUNTEERS

Community-level volunteers were selected from the communities—approximately one from every 25 households and trained with the help of the TOT BDRCS district level volunteers. A total of 1001 volunteers (Bagerhat 145, Baraguna 299, Patuakhali 207, and Pirojpur 350) were trained in 42 batches of 3-day trainings. These trained community level volunteers through participatory methods (explain in an appendix or dedicate a paragraph to describe) developed 93 psychosocial profiles (Bagerhat 12, Barguna 22, Patuakhali 30, and Pirojpur 29) in the 49 target communities in the four districts. The community level volunteers provided PFA to the assigned households in the stressful circumstances (as generally seen during the disillusionment phase of disasters) and day-to-day crisis. They also identify individuals requiring referral and, with the help of the

TOTs, facilitate in referrals to relevant sources of support (in this case, the general physicians trained by WHO at the district level).

The skills developed in the community volunteers were as follows:

1. *Assessment:* Assessment of the psychosocial profile of their communities (needs, interests, resources, and capacities with respect to the psychological and social aspects of the community).
2. *Information dissemination:* Providing credible information on psychological reactions to disaster and basic measures to manage at the household level. The trained community level volunteer provides information to the assigned households on psychological reactions and basic level interventions, self-care, information on issues concerning (stressor) the community or the specific household and distributes leaflets of psychological reactions and self-care explaining the content.
3. *PFA:* Providing PFA by the community level volunteers including identification and facilitation of referrals. Sharing meetings within communities are conducted that are directed towards ventilation of emotions and reduction of stress (including dissemination of information at community level and planning).
4. *Enhancing resilience through community activities:* Activities comprising a mix of psychologically designed activities and recreational activities to enhance psychological recovery of children and promote resilience for children every fortnight (at least 10 per village by project end) with the aim to involve majority of the children of a village in each activity.

8.4.3 ASSESSMENT (PSYCHOSOCIAL PROFILE) OF EACH COMMUNITY, INFORMATION DISSEMINATION, AND COMMUNITY PSYCHOSOCIAL RESILIENCE BUILDING

Guidelines were drafted for the trained CFs as how to conduct meetings, what topics to cover in which meeting, what games to be used for knowledge transfer, what technical information to disseminate, what information, education and communication (IEC) materials to use, and how to plan and execute resilience-building activities (Table 8.1). In addition, following Maldives program manuals for activities for children were developed. The resilience-building activities were similar in nature as in the Maldives PSP (see above) and children activities were also similar.

TABLE 8.1 Content of the Sharing Meetings Planned, Dash (2004)

Topics
Post-disaster psychological issues and symptoms
Identification of psychological issues within family
Accessing social support systems
Referrals and contact information for mental health care systems
How to help one's own self and family members
Identifying distress in children
Psychosocial support to children
Topics 1 and 2 in Sharing meeting I
Topics 3 and 4 in Sharing meeting II
Topics 5, 6 and 7 in Sharing meeting III
Topics 6 and 7 will also be shared in the children's activities (repeated in view of their complexity)

Sharing meetings within communities were planned to be conducted directed toward ventilation of emotions, reduction of stress, dissemination of information at the community level, and planning of activities. The program targeted to involve every individual at least three times within the program duration.

Activities were planned to be conducted for *children* every fortnight (at least 10 per village by project end). Its aim was to involve the majority of the children of a village in each activity (which comprised a mix of psychologically designed activities and recreational activities to enhance psychological recovery of children and promote resilience).

Social and cultural activities directed toward restoration of socio-cultural norms for enhancing recovery as well as strengthening of bonds within/between communities for better response to community crisis should be conducted. The included activities are celebrations of a culturally important day together by the community/communities and weekend fairs with the participation of various communities, etc. The program planned to support one major social event in each community within the program duration involving the majority of the community members.

Activities directed towards *addressing challenges in the community* such as restoring damaged community assets, community cleanup effort, renovation of community center/cyclone center, drawing up community crisis response plans, and so forth. It also includes any other concrete, purposeful, common interest activities that involve enhancing social cohesion within/between communities.

The community level volunteers conducted few activities in target communities, for example, 20 in Bagerhat, 20 in Barguna, and 38 in Patuakhali.

8.4.4   DISSEMINATION OF PROGRAM INFORMATION WITH EXTERNAL STAKEHOLDERS

A country-level coordination meeting was held in Dhaka on the RCRC PSP. The participants were personnel from BDRCS, International Federation, International Committee of the Red Cross (ICRC), German Red Cross, Danish Red Cross (regional office), the International Federation—reference center for psychosocial support, building community disaster preparedness capacity (BCDPC) Project and Department of Clinical Psychology (Dhaka University) as well as representatives from other organizations such as the WHO, the Danish International Development Agency (DANIDA), Australian aid, and Comprehensive Disaster Management Program (CDMP)/Government of Bangladesh. The meeting had the following objectives:

1. Update BDRCS senior management on PSP by the RCRC Movement, Asia and in the Sidr recovery program, Bangladesh
2. Share the PSP of the RCRC Movement in Sidr operations with other stakeholders in Bangladesh
3. Highlight areas of possible incorporation of PSP in Bangladesh as per the current international standards (Sphere and MHPSS)
4. Acknowledge the contributions of the department of clinical psychology, Dhaka University in the PSP in Sidr operations.

Such other small multiple meetings were conducted with other agencies disseminating information on psychosocial support programming such as ICRC, World Food Program (WFP), UNICEF, and Save the Children.

Unfortunately, the program could not be implemented beyond this point in view of management and liaison issues despite the availability of funds. Therefore, the impact of the program also could not be assessed as in the Maldives program. The design of the program had already incorporated many of the lessons learned from the Maldives program. There was more involvement of the stakeholders in the program design and quantification of the process through a number of activities conducted in community and number of contacts of individuals with program activities.

8.5   SUMMARY AND LESSONS LEARNT

Programming in the different countries clearly indicated that developing ideal programs, like a research design, in clinical studies is far from practicality. Managing the ever challenging realities on ground such as the existing governmental systems, other nongovernmental agencies working on ground with overlapping programs and liaison issues, management and policies of the parent humanitarian agency implementing the program, the target population, and the sociopolitical situation, always have strong impact on the programs making it impossible to maintain the program design and setting at perfection.

The reader can conclude that in the Odisha program the program duration was affected by time limits set by the implementing agency, given the fact that the program was the earliest of its kind and required a lot of innovations. In the Maldives program, human resources were scarce, cultural and linguistic contextualization difficult, a lot of technical development was needed and logistics was extremely challenging, to manage within the proposed 3-year timeline; and in the Bangladesh program, implementation challenges and extremely short duration of the program were the limiting factors.

The program design needs processing with planned changes in order to achieve possible changes in the field situation through programming over the years for smooth implementation. The lesson is that a PSP should be of 3 years or more. These community-driven programs call for a lot of innovation and it becomes extremely taxing on program managers and staff. The external stakeholder providing funds many times require immediate positive results. Implementation and financial limitations within the given time becomes a nightmare for the staff on the ground and programs end up with incomplete activities and poor fulfillment of monitoring and evaluation parameters.

Comparing the Odisha and the Maldives program there was a huge leap in programming with emergency phase, designed transition to long-term phase, planned activities in the communities and schools, preplanned program, exit strategy and evaluation of impact. The Bangladesh program took up the lessons from Maldives program by involving the target population in programming and implementation. Comparing the Odisha and Bangladesh program the reader can see that within a year of programming in Bangladesh, seven times the number of community volunteers trained in the Odisha program could be trained, whereas similar numbers of master trainers and district level volunteers trained. Also, the community level trainings were delivered by the trained TOTs of the immediate higher level in the training (Dhaka university students) and not by the program staff.

The experience, best practices, and lessons learnt in both the Maldives and the Odisha program—involvement of in-country human resources, communities, and local authorities in the implementation; designing programs as per the local context, culture and building on and adding to the existing systems and openness to liaison—was the major contributory factor for the speed and smooth implementation of the program. Unfortunately, since the program in Bangladesh had to be abruptly shut down, the program impact and monitoring and evaluation issues could not be assessed or fine-tuned, respectively.

In general, the programming in the three countries show how tiers of community and other in-country volunteers trained and involved in community, school activities, and TOTs helped the program unfold and reach out to the masses; how activities could be conducted at the community and school levels simultaneously with trained volunteers; and how the culturally and linguistically appropriate materials and dissemination activities helped acceptance, transfer, and sustainability of knowledge on psychosocial issues and programming within the stakeholders in the country of operation (see Jacobs, 2007).

To this day, monitoring and evaluation are major challenges in community-based psychosocial programming. Nevertheless, the third-generation program gives a distinct hint of advancement than its predecessors and suggests that further fine-tuning be done at technical advisors' level for future program implementation before the experience is lost and the generation of PSP begins again.

In addition, there are changes in the characteristics of sociopolitical situation of the target communities with the course of time following a disaster (see Table 8.2). In phase I, encompassing the traditional "Heroic" and the "Honeymoon" phases and covering roughly first 4–6 months of the disaster the primary focus would be on PFA to manage the direct psychological impact of the disaster, the losses itself and on the reduction of chaos and restoration of temporary order in the community. Phase II is from 4–6 months to roughly 1–1.5 years corresponding to the "disillusionment phase" and early part of the "restabilization" phase, and the focus is on management of stress due to secondary issues, need for accurate information, conflicts in communities, and community mobilization in the context of sudden social disruption. Phase III extends from 1–1.5 to 3–5 years corresponding to the "restabilization" phase and the focus is on the development of individual resilience, reestablishment of stable community structure in the context of disruption of the social structure of the community and development of state support and/or capacity for psychosocial support.

TABLE 8.2 Details of Psychosocial Support Training by Task, Setting, and Type of Training Developed for this Chapter. (*Source:* Satyabrata Dash, Personal Notes, 2003.)

Level	Task	Setting	American Red Cross Training
I	Establish contact, teach skills to affected people for taking care of themselves, by applying psychological first aid (PFA)	Neighborhood, village, community and emergency workers	Emotional support brigade (ESB)
II	Support family, friends, colleagues, and neighbors. Promote safe spaces for children	Neighborhood, community, schools	Community and school facilitators
III	Prepare to seek and receive assistance from external stakeholders (GO, INGO, local NGO)	Neighborhood, community and schools	Crisis intervention technicians
IV	Provide assistance to affected people in the form of: homecare, social work, general practitioner (GP) physicians, and reestablishment of place	Community center, school, health post	Crisis intervention specialist
V	Medical and psychiatric treatment	Hospital	None available

GO: Governmental organizations, *INGO:* International nongovernmental organizations, *NGO:* Nongovernmental organizations

Program developers must take this into account while setting up the program design and monitoring the indicators or else the activities designed for encompassing a particular phase will be irrelevant later. The program materials have to be linguistically and culturally appropriate and for acceptability of the target population involvement of the target population in designing of program, activities are important. The conviction of the local authorities on the PSPs is a key requirement for implementation of the program. The liaison with governmental and nongovernmental agencies and avoiding duplicity of program components with other agencies in a post-disaster scenario is a challenge (Fig. 8.3).

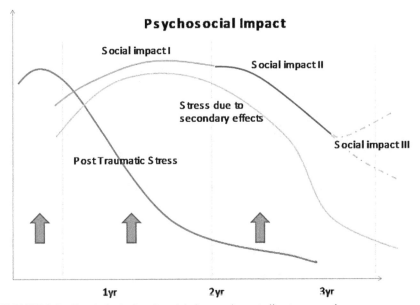

FIGURE 8.3   Psychological and social changes in post-disaster scenarios.

Source: From S. Dash, PostDisaster Psychosocial Support: A framework from lessons learnt through programmes in SouthAsia, The Australasian Journal of Disaster and Trauma Studies, © 2009 by S. Dash, Reproduced with permission.

Therefore, before disaster agencies need to employ technical advisors to design and develop PSPs for different situations as per the agency mandate from the innumerable resources developed by various agencies all over the world till date. A bright example of this is the IFRC's Psychosocial Support Centre. When disaster strikes, these technical managers are expected to come into the country of operation and together with the country program manager contextualize the program to the country and situation and then allow the country program team to carry out the program and help guide the team, when required, until program completion and final evaluation is done. All lessons learned and best practices are to be preserved for future programming. The MHPSS guidelines by the Interagency Standing Committee in 2007 is also another great example where all the experiences, science, and planning have been put into one document for assistance in psychosocial support programming.

8.6 POSTSCRIPT

It is rare to find a psychiatrist working with community members to enhance psychological well-being. After active participation in the development of these three programs, and as a social psychiatrist my contributions in the development of the program, were informed by the following:

1. Understanding the perception and feelings that people had about their communities, ways of life (cultural issues) and thinking patterns to develop contextualized programs and materials to make it acceptable to them.
2. Identifying the goals of the program, from the eyes and ideas of community members and steps to reach them through program activities.
3. Providing the scientific knowledge on the program designing, activities, and monitoring and evaluation.
4. Analyzing the overall program outcome—the strengths to replicate and the flaws and methods to set right in future programming.
5. Dissemination in the appropriate global forums for the scientific world to know and use the experience.

8.7 ACKNOWLEDGMENT

I, hereby, acknowledge the program documents—proposals, reports, and evaluation—of the Indian Red Cross—ARC program in Odisha, India; ARC program in Maldives and IFRC and Red Crescent Program in Bangladesh as a source of data for the above chapter.

REFERENCES

American Red Cross Program Documents.

Dash, S. (2003). Disaster Mental Health Psychosocial Care Orissa: The Safe Homes Program Proposal . Indian Red Cross Society, Orissa State Branch (unpublished paper).

Dash, S. (2004) Disaster Mental Health and Psychosocial Care Evaluation. Indian Red Cross Society, Orissa State Branch (unpublished paper).

Dash, S. Psychosocial Impacts of Disaster and Program Initiatives in Asia. *Asian Disaster Management News.* Jan-Mar (12–13) 2006.

Dash, S. Post-Disaster Psychosocial Support: A Framework from Lessons Learnt Through Programs in South-Asia. *Australas. J. Disaster Trauma Stud.* **2009,** *2009*(1), 21–30.

Dash, S.; Khoja; Das. 2003.

Dodge, G. Interview for Report Entitled "Indian Govt Under Fire for Neglect of Injured Quake Victims". Agence France-Presse. 2001, (accessed Feb 15, 2001).

Gerard, A.; Jacobs G. A. The Development and Maturation of Humanitarian Psychology. *Am. Psychol.* **2007,** *62*(8), 929–41

Gupta, K. Disaster Management and India: Responding Internally and Simultaneously in Neighboring Countries. In *Comparative Emergency Management: Understanding Disaster Policies, Organizations, and Initiatives from Around the World;* McIntyre, D., Ed.; Emergency Management Institute: Emitszburg, PA, 2006.

Indian Red Cross. Program Monthly Reports and Final External Evaluation, 2004.

Jacobs, G. *Jacobs will Oversee Mental Health Program in the Wake of Indian Earthquake.* Plain Talk: University of South Dakota, April 6, 2001. . http://rop.plaintalk.net/cms/news/story.html. (accessed March 31, 2016).

Mathews, W. *The Psychosocial Support of the American Red Cross International Services;* American Red Cross International Services: Washington, D.C., 2002.

Prewitt Diaz, J. O.; Biel, D. *Organización de un programa psicosocial entre entitades civicas y sociedad civil despues de un deslave en Nicaragua*; Cruz Roja Nicaragüense y Cruz Roja Española: Managua, Nicaragua, 2002.

Prewitt Diaz, J. O.; Dash, S.; Das, M. *A Proposal for the Implementation of a DMHPC Program in Four District of Orissa Affected by the supper cyclone*; Indian Red Cross Society: Bubhenehwar, Orissa, 2003.

Prewitt Diaz, J. O., Lakshmlnarayana, R.; Bordoloi, S. Psychological Support in Events of Mass Destruction: Challenges and Lessons. *Econ. Polit. Wkly.* **2004,** *39*(21), 2121–2127.

PART III

Current Events That Require Psychosocial Support

PSYCHOSOCIAL SUPPORT AS A PRIORITY IN EBOLA RESPONSE: STRUCTURED INTERVENTIONS IN ACCORDANCE WITH IASC-MHPSS GUIDELINES

PRAMUDITH D. RUPASINGHE

United Nations Mission in Liberia, Liberia, West Africa;
pramudith.rupasinghe@icloud.com

CONTENTS

9.1 INTRODUCTION

The Ebola crisis profoundly impacted the psychosocial well-being of the population of three majorly affected countries, that is, Liberia, Sierra-Leone, and Guinea because of different preconditions such as extreme poverty, insufficient health facilities, and mental health and psychosocial issues due to civil conflicts resulting in loss of lives, loss of education, widespread fear, mistrust, loss of livelihood, and inability to practice social and cultural practices (World Bank, 2014).

The population that was directly affected by the epidemic manifested diverse challenges to social reintegration including relational barriers, distress, anxiety, grief, and fear. In addition, this document reflects on the community-based coping strategies and cultural practices that were successfully adapted by the affected communities in the face of adversity.

Thus, as the key learning from the response, many of the emergency response actors involved in the response that the mental health and psychosocial support for those affected by the humanitarian crisis should be centered on resources, strengths and culture, build on existing capacities at the local level and delivered through the local structures and the channels that the local population has relied upon for decades. As illustrated in the intervention pyramid (see Fig. 9.1), it became clear that there was a need to ensure that from the onset of the crisis, linked multilevel supports must be made available. The first layer, that is, basic security and services, represents the emergency response required to protect the psychosocial health of the entire population.

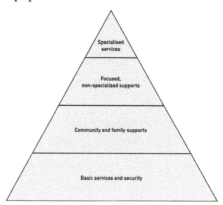

FIGURE 9.1 Inter-Agency Standing Committee Guidelines on Mental Health and Psychosocial Support Intervention pyramid.

Four-layered psychosocial intervention pyramid approach to the Mental Health and Psychosocial Support (MHPSS) guidelines of the Inter-Agency Standing Committee (IASC) was widely utilized during all phases of the response. *Basic services and security* which is the foundation layer was one of the main interventions first delivered to the communities at the entry phases. "There was a lot of resistance and denial of health workers, yet there were a lot of needs hidden under the defense walls of denial and resistance," Sedo, a community volunteer from Guinea added. By enabling the access to the basic services for the communities that were affected by the epidemic, the community health workers and community mobilizers/volunteers could access the communities and gain the trust to reach to the second layer of the intervention pyramid, that is, *community and family support* which remained vital throughout the epidemic. The daily struggles of communities were: knowing about the others enrolled in the Ebola treating centers, emotional support for grieving families, and accepting that the life has changed. *Community and family support* was extremely important for such cases (IASC, 2015).

A smaller subset of the population affected by crises will be able to maintain their psychosocial well-being if they receive help in accessing *community and family supports*, the second layer of the pyramid, which are often disrupted by crises or emergencies. These supports might include family tracing and reunification, memorials, parenting groups, and the activation of social networks, such as women's groups and youth clubs.

Psychosocial first aid (PFA) and basic mental health care by primary health workers comprise the pyramid's third layer and are necessary for the still smaller number of people who require additional and more focused individual, family, or group interventions by trained and supervised workers. For example, survivors of gender-based violence might need a mixture of emotional and livelihood support from community workers. The global expert guidance indicates that 10 15% of affected people require *specialized services*, the pyramid's final layer, such as professional psychological or psychiatric support.

A recent analysis found that although the clear majority of mental health and psychosocial support programming in humanitarian settings focuses on strengthening family- and community-based supports, existing research focuses almost exclusively on the top layer of specialized clinical mental health services. The risk of causing harm by continuing to implement untested interventions is significant. The lack of a solid evidence base is one factor in the continued implementation of ineffective programming, for example, psychological debriefing, which has been found to exacerbate symptoms of distress.

9.2 PSYCHOSOCIAL SUPPORT AS KEY TO POST-EBOLA INTERVENTIONS

As of 2016, there are a total of 17,423 Ebola survivors in three most affected countries in West Africa, that is, 5868 in Liberia, 10,106 in Sierra Leone, and 1271 in Guinea. In all 28,544 people in these countries (4809 in Liberia, 3955 in Sierra Leone, and 2535 in Guinea), total 11,299 have died leaving their loved ones behind, and many of them have experienced multiple bereavements (Wikimedia Foundation). UNICEF estimates that there are over 16,000 orphans because of Ebola in affected countries with a significant number of orphans having lost both parents or guardians.

Diverse efforts, such as setting up Ebola treatment units (ETUs), rapid training of community members on surveillance, community interventions, strengthening the government health, providing psychosocial support to the affected population, have been made by the nongovernmental organizations, UN agencies as well as government institutions. This has led to the eradication of the disease in Liberia, Sierra Leone, and Guinea. However, the effects of Ebola will continue to be felt for a decade not only in current generation but also in the future generations. In such case, mental health psychosocial support has become paramount after the end of the crisis as most of the issues left behind, unseen, and unnoticed are psychosocial issues.

Besides addressing the needs arising from the potential long-term health-related consequences of the disease that many are concerned about, the psychosocial consequences, such as loss, grief, shame, distress, and suffering, should be taken into consideration in designing post-Ebola interventions because psychosocial impact of Ebola is relatively higher than other epidemics because of the myths, misperceptions, and stigma linked to the disease because of its deadly nature. Thus, usually, the survivors and their family members tend to become victims of stigma and discrimination which worsen the psychological impact of loss, death, and grief caused by the disease itself. The Ebola outbreak has resulted in a wide range of psychosocial protection concerns experienced at the individual, family, community, and societal levels in an extreme magnitude than in case of any other natural disaster. Over the course of the outbreak, normally protective supports such as school, work, basic preventative health, community groups, and daily routines were disrupted while preexisting problems of social injustice and inequality were amplified, especially in Liberia and Sierra Leone being countries that have been severely affected by civil conflicts.

As could be expected in an emergency, a significant number of individuals in the affected countries reported a range of psychosocial difficulties associated with, or amplified by, the emergency and the response. Majority of people affected at diverse levels reported experiences of complex grief associated with multiple losses.

The specific context of Ebola and the problems that come with it (e.g., the fear of infection, inability to care for loved ones, the shock caused by transportation to treatment centers, witnessing deaths, and culturally inappropriate burial practices) complicated and, in most of the cases, increased people's experiences of bereavement and loss (Szumilas et al., 2010). Stigmatization and discrimination of survivors and their family members had left diverse hurdles for social reintegration making the situation more complicated.

Dynamics of interpersonal relationships changed significantly during the outbreak, interpersonal conflicts based on distrust and blame associated with specific events during the emergency rather than a person's general status as a "survivor." In many cases, an individual or family may blame another family member, friend, or a community member for bringing Ebola to the community, infecting a loved one, or calling emergency Ebola hotlines to report a suspected Ebola case. Even though Ebola is a natural disaster, people started blaming each other which was a normal reaction to an abnormal adverse event. Breakdown of the relationships at every layer created a challenging environment not only for fighting the epidemic but also for working with communities than the concept of "stigma" would suggest and required a nonconventional kind of response focused more on societal dynamics. Rather than conventional community-based sensitization on Ebola, more focused and participatory approaches were needed based on conflict resolution and mediation methods so that trust and relationships could be rebuilt.

Adults often denied that the children were encountering situations of discrimination whereas the children verbalized, in many occasions, that they were subjects of discrimination as a significant percentage of children who were Ebola survivors or orphans due to Ebola virus disease (EVD) related deaths of parents or guardians manifested feeling of being isolated in their peer groups and inclined to socialize with other children who had gone through the same experience as survivors or orphans creating a cub-group among the children that was noticeably marginalized.

Once the schools restarted in the affected countries, the relationships between peers, teachers, and the parents of the peers were not the same for the survivors or the orphans; relationships at each layer had deteriorated indicating that post-Ebola psychosocial interventions at peer–group and teacher–student settings were urgently required. It was challenging for the orphans as well as guardians to adapt to the new family structures. In some places, the issues such as faith and tribe had come between the guardians and orphans, creating a complex situation where negotiation and mediation had a significant role besides psychosocial interventions.

Apart from that, guardians, most of whom were women, had experienced one or more losses of family members by Ebola and were caring for between one and ten additional children. Some children and their guardians reported shortages of food, bedding, and funds for school fees (Inter-Agency Standing Committee, 2007).

Amidst of the financial difficulties and social barriers that the guardians were facing, some orphaned children had felt they were treated differently in comparison to the biological children of the guardians. However, on the other hand, guardians encountered diverse difficulties in providing for children economically and emotionally, particularly as children were grieving and in certain cases were distressed, withdrawn, or manifesting mood swings or emotional difficulties.

Continuous family-focused support, with special focus on guardians and orphans, is a must to support the development of health-sustaining relationships, improved psychosocial well-being, and enhanced residence and sense of place. Nonetheless, the establishment of a sense of place among the orphans as well as the guardians demand a constant multitiered psychosocial intervention that covers all layers of MHPSS intervention pyramid for a longer period which is one of the most challenging yet vital needs in post-Ebola context.

9.3 PSYCHOSOCIAL CHALLENGES ENCOUNTERED BY EBOLA RESPONSE TEAMS

Many humanitarian workers, especially health and social workers, had noticed that explicitly targeting survivors for support causes resentment and could hinder reintegration. Global evidence (Ebola Virus Cases in Guinea, Liberia, and Sierra Leone, 2014) supports service provision based on identified needs rather than crude categories such as "Ebola survivors" or "Ebola orphans." Specialized support should also be provided for workers who experience distress because of aiding people affected by Ebola.

High levels of distress and anxiety have been experienced by families who did not receive timely and accurate information about the status or whereabouts of their loved ones. Sometimes the communication gaps allowed the rumors to take the lead in community actions resulting community resistance for surveillance, safe burials, taking the patients to the Ebola treatment units, and implement other supportive activities in the communities (Centers for Disease Control and Prevention, 2014).

At the peak of the crisis from August to October 2014, in many places, especially in Guinea and Sierra Leone, family members did not receive information regarding the whereabouts of their loved ones who were taken by ambulance for treatment. They received conflicting information from different service providers and did not know whether their loved ones were still alive or not. In such cases, the myths, such as foreign agencies, take people for extracting organs and blood leading to severe attacks on humanitarian workers who were on the ground (Fu et al., 2015).

Furthermore, the safe burial practices were perceived as disrespectful and harmful by many community members during the early stages of the outbreak and caused a lot of spiritual distress among the affected populations in Liberia, Guinea, and Sierra Leone. Many community members highlighted witnessing disrespectful or inappropriate burial practices and in some cases, these experiences continued to cause them severe distress.

Apart from the above-mentioned challenges, local capacity in terms of specialized mental health services was limited. In Liberia, there was only one local psychiatrist for the whole nation of 4.3 million and there were no trained local psychologists. In Guinea and Sierra Leone, similar conditions existed which limited specialized services. Simultaneously, due to the stigma toward the mental disorders locals tend to avoid using available mental health services because the person suffering from such disorders is widely called "mad" irrespective of the nature or intensity of the problem that the individual faces.

9.4 COMMUNITY-BASED MECHANISMS AND STRATEGIES FOR ESTABLISHMENT OF SENSE OF PLACE AND ENHANCED RESILIENCE

As a result of disruption of the daily routine and limited social interaction, commodities became inaccessible, usual livelihood became impossible, and the schools and other public places were shut down; a sense of safety, community efficacy, connectedness, hope and ability to cope with the crises became almost impaired. In the three most affected countries discussed earlier, sense of place was highly threatened.

Making practical attempts to establish a spiritual sense of place and cultural sense of related activities, such as engaging in religious, spiritual, or cultural actives that the community members believed, reduced the risk of infection. Also, cultural narratives of risk were in the nucleus of health and it was vitally important that they are taken seriously by those attempting to control the spread of communicable diseases.

Reestablishment of sense of place in terms of security or feeling safe in terms of the probability of getting infected during the outbreak by taking control of infection and by practicing recommended infection-control measures (hand-washing, avoiding body contact, etc.) had surfaced in a different shape unlike in the other disasters like earthquakes or tsunami. The community members by themselves followed strategies that were most appropriate to the local context while meeting the standards of infection-control measures set in place by the humanitarian actors as well as local health authorities (Snow, 2014).

Community participation in the establishment of a sense of place in the most affected communities and in public that was also in panic and experiencing severe level descriptions in daily routine due to extreme fear, immensely helped in Liberian context. Clergy of the two most widespread religions in the affected countries—Islam and Christianity—contributed in facilitating the communities in establishing a spiritual sense of place while raising the level of hope among the hard-hit population. In addition, many churches and mosques supported to bridge the gap between health workers and the community members at a time when the truth and the hope had been shattered by the rumors and the myths. Cultural practices such as singing and praying together brought a sense of coherence and optimism in many teenage and older survivors whereas many bereaved individuals gave a sense of hope to their close family members and neighbors in the time of fear and despair.

The human capital sense of place in many situations was quickly defragmented. Especially in Liberia, identification of risk and resources in the affected communities, which was done in a gender-, faith-, and age-sensitive manner, allowed the sensitization operations to be carried out easily and also helped in the mobilization of the community members effectively.

Utilizing social mapping, children and adults mapped their relationships and showed the dynamics of relationships during the outbreak. The boys and girls who were isolated by their peers reported their attempts to repair and restore their friendships, though these efforts were not always successful. Women and men who had guardians of children reported their attempts to create a safe, secure family context, in the face of real difficulty by providing emotional support when their own resources were depleted by multiple losses. Therefore, when children are under guardianships, their sense of place is impacted. At the same time, many community members in Guinea, Sierra Leone, and Liberia showed a tendency of involving in activities such as supporting others. For example, one teenage girl who had survived Ebola and lost her parents to the disease talked about the sense of well-being and purpose she acquired from volunteering to support the children in the orphanage. A significant number of survivors worked in the Ebola treatment centers lately. As per them, the work they do in the treatment center helped them to gain a sense of worth by being productive and efficient and uplifted their self-esteem while contributing to the fight with Ebola.

Therefore, it was visible that establishment of sense of place in its five dimensions—cultural sense of place, ecological sense of place, spiritual sense of place, human capital sense of place and problem-solving, and osculation oriented sense of place—should be paramount for future psychosocial program development in a highly participatory way to prevent undermining the important process of self-efficiency and community efficacy by fully utilizing the resources and capacities available in the affected population, treating them as dignified and capable human beings.

There were instances the medication such as sleeping tablets, Paracetamol, or benzodiazepines were provided to community members when they became overwhelmed due to their losses (NIH, 2014). In Sierra Leone and Liberia prescribed medication is used to recover from the shock or emotional reactions after several months of their bereavement. This behavior suggests an unmet need for psychosocial support with calming activities, relaxation techniques, and sleeping. Usually it is noticed in Sierra Leone and Liberia that the prescribed medication is used to recover from the shocks but long-term use of these even after several months of their bereavement which is unusual clearly indicates an unmet need for support with calming, relaxation, and sleep. However, many programs followed the cognitive techniques as well as simple exercises that help in relaxation.

However, trained volunteers, who were deployed in the affected areas, helped significantly to overcome the stigma-induced anxiety and stress reactions of survivors, their family members, and the family members of people deceased with Ebola by stepping up the communication and providing lifesaving information to communities so they better understand the disease and know how to protect themselves. The affected families must get proper treatment, but after they have recovered or are declared free from the virus, they need the community's support to return to normal life. Relaxation techniques such as breathing exercises, music, plays, and traditional dancing programs; practical skills and stress management for dealing with acute anxiety and stress, as per the "focused nonspecialized support" in IASC-MHPSS guidelines, should be implemented to help people recover.

However, in many treatment units, there were counselors and psychologists assigned to provide more focused services for addressing the need for specialized services, especially for the survivors, making sure that the whole response was structured within IASC-MHPSS thematic framework.

9.5   FIGHTING STIGMA IN POST-EBOLA PHRASE; IMPLICATION OF IASC-MHPSS GUIDELINES

As Ebola is a highly contagious disease that spreads very quickly and has no cure, the most common human reactions are fear and stigma. Most often stigma and fear are interrelated in case of Ebola; stigma is a reaction to fear. Therefore, the daily routine of life was heavily impaired in affected countries because of stigma and fear. It was noticed in many places in Uganda and the Democratic Republic of Congo, as a common reaction, that the people limited their movements and refused to venture too far from their residences when Ebola epidemic hit. Especially the family members of the individuals deceased due to Ebola or the survivors had to go through stigma and they were often perceived as fear inducers. The survivors, their family members, and the family members of deceased individuals were deprived of their social interactions, sometimes including their right to schooling, right to travel in public transportation, etc. There were also instances that the survivors, their family members, and the relatives of the deceased were thrown out from the rented dwellings.

"In West Africa people have never experienced anything like this in recent past to which the community memory can be associated" Tamba, a community leader in Lofa said. The fact that Ebola is perceived as a new disease in this region which is highly infectious and contagious also contributes to the fear and stigma attached to it. Crossing all borders from its epicenter in less than 5 months, Ebola has intruded three West African nations while it started spreading at a condensable extent in neigh-boring countries of affected nations (Relief Web, 2014). When cases were reported from Mali, Senegal, and Nigeria, the authorities of respective countries decided to close the borders and restrict the movements from and to the affected countries (NPR, 2014). Moreover, most of the airlines stopped coming to Liberia, Sierra Leone, and Guinea. "It was as a measure of precaution but the commoner's viewpoint from the rest of the world started adapting to such a level that they saw the West Africans as viruses." Joseph B K Kamara—Psychologist/Staff Counselor of UN mission in Liberia added. While stigma and fear crossed the borders of affected countries, stigma inbound intensified. "When I returned from ETU, my rented room was already given to someone else, many of my belongings were burnt; the landlord told me that he did not have rooms for me and he closed his door behind" John, who was a survivor from Liberia, told a part of his survival story as an Ebola survivor in the society. Survivors of Ebola whose family members died also suffered from stigma. Even after the recovery, the community still believed they have contracted the Ebola virus and did not want them in the markets, in their houses, or places of worship. In many cases, this hindered the healing process of psychological trauma that they had gone through. Avoiding direct contact with people infected with Ebola virus is one of the major measures used to minimize the spread of the disease. However, this also had a negative effect as people who suffered from other severe illnesses, such as malaria, were sometimes admitted into isolation as a precaution, and when they recovered and were discharged, the community still believed they were being treated for Ebola and could still be contagious. Fear of being marginalized or isolated also caused people to conceal their illness. Hence, fight against stigma was vital for ensuring a dignified social and personal life not only to the survivors but also to their family members and close contacts.

Continuous sensitization programs were conducted by diverse actors who were part of the response, including Minister of Health (MoH) of all three affected countries, Red Cross and its partners, and UNMEER—United Nations Mission for Ebola Emergency Response. Identifying and tracking those who had come into contact with suspected cases, disinfecting the homes of Ebola victims, and raising awareness among communities on how to protect themselves from becoming infected without being discriminative to the survivors, patients, or their relatives helped in changing the common perception at the later part of the epidemic (Centers for Disease Control and Prevention, 2015). Being supportive without discrimination was one of the key messages given by the community sensitization teams. Furthermore, a certificate was issued to the survivors upon being discharged from the ETU which helped them to get back to society as normal and accepted individuals. However, social reintegration of survivors still remains challenging in all three affected countries. At the same time, the potentiality of a survivor to transmit the virus via sexual intercourse even after the recovery (till 90 days) and the isolated cases reported because of sexual contact with the survivors fueled the remaining fears and the stigma.

9.6 FACILITATING COMMUNITY RESILIENCE THROUGH LONG-TERM COMMUNITY OWNED PROGRAMMING INSTEAD OF MERE RESPONSE

It was evident throughout the response that long-term and appropriately resourced psychosocial programs were required to address the needs identified at the local level with the active participation of local population.

Community-based psychosocial programs founded on local beliefs and sensitive to the social, cultural, and religious aspects of affected countries remained vital need. In addition, training as a long-term national-level capacity building strategy along with clinical supervision was highlighted in Liberia, Guinea, and Sierra Leone.

Short-term project-based funding is not sufficient to meet the identified psychosocial needs in Ebola context as many of the needs required long-term social intervention in post-Ebola phase. Without adequate support, actions to lessen the impact of the crisis on people's lives, livelihoods at times aggravate the condition while long-term physical an. Appropriate, community-based programming designed on the basis of IASC-MHPSS, focusing on community resilience will prevent escalating of problems and strain clinical settings. As per IASC-MHPSS, post-Ebola interventions should not target merely the affected families or the survivors but the community channeling through the existing familial, social, and cultural settings, as a strategy of holistic community resilience and to prepare communities for living with Ebola while facilitating the recovery. Effective interventions will enhance naturally occurring supports and mobilize local resources and strategies that foster well-being and bring about resilience at family and community levels. This can involve linking to existing family and community support, focused, nonspecialized supports and specialized services as indicated in the IASC-MHPSS interventions triangle. Strong referral mechanisms within and between layers are vital. Apart from that, the psychosocial programs should follow an intersectional approach that involves health care, education, and protection spheres. Therefore, rather than imposing predetermined preplanned intervention methodologies in supporting people affected by the epidemic, it became vital to pay proximate attention and to refrain from undermining already existing coping mechanism and community-owned patterns of resilience.

REFERENCES

Centers for Disease Control and Prevention. Ebola Epidemic—Liberia, March–October 2014. Centers for Disease Control and Prevention, 14 Nov. 2014. https://www.cdc.gov/ (accessed April 9, 2016).

Centers for Disease Control and Prevention. Ebola Outbreak—2014 Centers for Disease Control and Prevention, 15 Oct 2015. http://www.cdc.gov/media/dpk/2014/dpk-ebola-outbreak.html (accessed April 29, 2016).

Ebola Response Anthropology Platform. Stigma and Ebola: An Anthropological Approach to Understanding and Addressing Stigma Operationally in the Ebola Response. Policy Briefing Note. Dec 11, 2014; pp 1–5. http://www.ebola-anthropology.net/wp-content/uploads/2014/12/Stigma-and-Ebola-policy-brief-Ebola-Anthropology-Response-Platform.pdf

Ebola Virus Cases in Guinea, Liberia, and Sierra Leone, as of July 20th. *Afr. Res. Bull.: Polit. Soc. Cultural Ser.* **2014,** *51*(7). http://www.who.int/crs/don/2014_07_15_ebola/en

Fu, C.; Roberton, T.; Burnham, G. Community-Based Social Mobilization and Communications Strategies Utilized in the 2014 West Africa Ebola Outbreak. *Ann. Glob. Health* **2015,** *81*(1), 126–127.

IASC Reference Group on Mental Health and Psychosocial Support in Emergency Settings. Mental Health and Psychosocial Support in Ebola Virus Disease Outbreaks. 2015; pp 1–32.

Inter-Agency Standing Committee. *IASC Guidelines on Mental Health and Psychosocial Support in Emergency Settings*; IASC: Geneva; 2007.

NIH Intramural Research Program. NIH Ebola Response. 2014. http://irp.nih.gov/catalyst/v22i6/nih-ebola-response (accessed April 30, 2016).

NPR. Borders Closure for Ebola Countries. 2014. http://www.npr.org/2014/08/23/342652013/borders-close-as-ebola-spreads-in-west-africa.

Relief Web. Psychosocial Support during an Outbreak of Ebola Virus Disease. 2014. http://reliefweb.int/report/world/psychosocial-support-during-outbreak-ebola-virus-disease (accessed Aug 28, 2015).

Snow, S. S. Environmental Infection Control for Ebola: Approach for a UV Claim. April 11, 2014.; pp 1–5.

Szumilas, M.; Wei, Y.; Kutcher, S. Psychological Debriefing in Schools. *Can. Med. Assoc. J.* **2010,** *182*(9), 883–884.

Wikimedia Foundation. West African Ebola Virus Epidemic. Wikipedia. https://en.wikipedia.org/wiki/West_African_Ebola_virus_epidemic. (accessed April 18, 2016).

World Bank. Transcript of Remarks at the Event: Impact of the Ebola Crisis: A Perspective from the Countries. October 9, 2014 (accessed April 28, 2016).

CHAPTER 10

THE FIRST PSYCHOSOCIAL PROCEDURES IN ZENIE

CECILIE ALESSANDRI[1*] and JEAN-CLAUDE KÉKOURA ZOUMANIGUI[2]

[1]*French Red Cross, Paris, France*
[2]*French Red Cross, Macenta, Republic of Guinea*
Corresponding author: Cecilie.Alessandri@croix-rouge.fr

CONTENTS

10.1 INTRODUCTION

In *February 2016*, the *French Red Cross (FRC)* psychosocial team in Macenta, Guinea assessed the psychosocial needs of the populations of Zenie village, in the Fassankony sub-prefecture at the border with Liberia. This village has been particularly hit by the *Ebola* hemorrhagic fever outbreak with 118 deaths registered for a population of approximately 1600 inhabitants. The psychosocial needs turned out to be particularly considerable in this situation. Indeed, many families have been partially or entirely decimated, which resulted in a general apathy of survivors and a disintegration of the social, economic, and family life. The last victim of the village died at the Gueckedou Ebola treatment center (ETC) on October 7th, 2014.

Moreover, with the needs still being expressed many months after the end of the crisis, the FRC psychosocial support teams have implemented psychosocial activities to encourage verbal expression while reestablishing the community and family solidarity and enabled the population to speak about traumas experienced.

During the first visit, a meeting took place with the village and new authorities to explain the psychosocial activities. Then, a meeting with the whole community was held at the village cultural center to present various activities and their objectives. This first approach enabled to sound out around 200 people. Further to the information meetings, the FRC/Macenta team managed to intervene and propose psychosocial activities, mainly group discussions, which were undertaken by psychosocial agents working in pairs.

The team gathered testimonies from inhabitants in smaller groups which separately included women, men, elderly, young people, and teachers; this enabled the teams to better apprehend the needs of the various groups and to identify the most affected individuals via observation and active listening. Among these people, the majority lost a relative infected by Ebola.

10.2   THE IMPACT OF THE EBOLA CRISIS ON THE POPULATION

10.2.1   THE IMPORTANCE OF HUMAN LOSSES: LONELINESS, HELPLESSNESS, AND TRAUMA

The teams were struck by the magnitude of human losses within the community. Everyone had lost at least one relative during the crisis. Thus, some houses were closed or the unit of a family reduced. The villagers have difficulty overcoming the horror scenes that they encountered during the peak of the illness and they are still very affected by them. A man in his forties said, trying to suppress his tears: "We sometimes registered more than ten deaths in the same day and people cried all over. I shall never forget this dark period of my village history." He added: "We thought that we were all going to die. Bodies were collected every day from houses."

These huge numbers of deaths put all families in a process of bereavement, as mentioned by this advisor of the village's new authorities: "We are all bereaved in this village. Each person lost a relative during this crisis and we are still affected." Therefore, due to bereavement, the village population lost the desire to continue to live "normally."

Indeed, all the people interviewed mentioned lack of courage to undertake or resume an economic activity. A lot of them decided to go and work in the field in the morning, but they quickly returned home for lack of courage or energy to accomplish their work. Some of them came back home crying, thinking of their deceased relatives: "my heart is broken when I go to the field while thinking of my lost relatives with whom I used to work. I start crying and I go back home straight after" a woman in her forties said, crying. "I still cannot work properly since the death of my relatives. I couldn't grow any peanut or rice in the field last year" another woman of the same age said.

Moreover, local beliefs have also an impact on the villagers, for example, some of them dreaded to encounter the ghost of their deceased relatives once they arrive at the field. A man of about 50 years old said:

"One day, coming back from a walk, I told my neighbors that I was sad to go back home since my wife was no longer there; she always used to keep something for me to eat. When I arrived home that day, I saw my wife's ghost. The same thing happened to me another day while I was alone in my field. Since then, I am afraid to go work in the field but I survived, thanks to the charity of the people willing to help me. I also do not wish to see the objects which belonged to my wife."

Therefore, the crisis had weakened the desire to undertake and build future projects, the population being unable to overcome the gap left by the deceased relatives and to accept the reality of the day-to-day existence which had become increasingly difficult.

The blow undergone by the villagers was huge and most people expressed discouragement or a persistent feeling of helplessness.

While listening to the villagers, the psychosocial teams noted that beyond human losses, the population thought that they did not have the resources to rebuild their own lives and overcome the situation. During interviews, many people showed great distress through cries, and the individuals who could suggest some measures to get out of the crisis were quite rare. They stated they felt helpless and incapable: "See, you should help us, we lost everything due to this outbreak, we do not have any resources today which might help us to recover," said a fortyish-year-old woman, following other women who complained of the same thing.

The people met have also insisted on the fact that this situation and the result in human losses are felt as a trauma affecting their psychological well-being. For the more vulnerable people, the trauma created a significant psychological unbalance, as mentioned by this fortyish-year-old woman: "Since the death of my brothers and sisters further to this epidemic, my mother, who is very old, feels lost and is almost crazy. There is no consistency in her speech and her behavior. Her conscience has been quite altered."

10.2.2   FEELING OF ISOLATION

With the *social and family* point of view, the situation in Zenie, created by the Ebola outbreak, was disastrous. It led to a deterioration of the social fabric such as the dissolution of associations and other groups, the breakup of certain family ties and the isolation of some groups of people.

One of the consequences of this situation was the isolation felt by some members of the community, especially the elderly, who had no assistance whatsoever, or some men and women who were in a single-parent situation. A man in his fifties said: "I have to prepare the meals for me and my children who are still young. This illness snatched away all those who took care of me." Another woman said: "I am old, I can no longer work in the field and those who supported me are all dead, I have no one to rely on, I am asking for your help."

The psychosocial teams noticed that the feeling of loneliness generated by the human losses sometimes lead to a risk behavior. A man in his thirties said: "I still remember my sister who was my whole life and it saddens me a lot. That is why I started to drink alcohol to feel better."

This isolation may also be related to certain difficulties faced by people who have withdrawn into themselves instead of trying to keep socializing. This self-exclusion limits the possibilities of the people who are suffering to express themselves and manage these difficulties better. A woman of about fifty said: "I do not wish to socialize with the others, we all share the same story and that's what comes up when we meet. I personally find that this often increases my moral suffering."

10.2.3 THE PROBLEM OF ORPHANED CHILDREN

The situation of the village poses a big problem to many children who became orphans due to Ebola. A total of 132 children from Zenie lost at least one parent and are encountering considerable problems. This large number of orphans can also be explained by the local traditions that enable one man to marry three women, each of them bearing five children. The man's death leads, therefore, to a significant number of orphans.

It is hard for those children to find spaces of expression that facilitate their bereavement process and the management of their suffering, further to the loss of one or both parents. The tensed environment and the emotional charge which should be dealt with by adults did not provide an efficient support to those children.

Children who have lost one of their parents are being taken care of by the survivor parent. When it happens to be the father many feel helpless and incapable of handling housework that is normally attributed to women by this community; or, when the deceased parent is the father, mothers encounter difficulties gathering the necessary resources to provide for the family.

When the children have lost both parents, they are either sent to orphanages which often lack the means to answer all their needs, or they are left under the care of a relative who ends up having one or more extra children to provide for and who do not have time to spare for new members of their family.

Some people of Zenie tried to provide support for those children, such as a 50-year-old teacher who met during the psychosocial sessions and opened a reception and education center for Ebola-orphaned children of the village. With his own financial funds and other means that he could find, the man received 72 orphans, every day, between the ages of 3 and 7. He mentioned, to the teams, the difficulties he was facing and the fact that he thought his action was crucial, saying that he was fulfilling his duty by fighting the collateral damages incurred by the crisis. He said: "since I opened this center, I haven't received any kind of support. I do whatever I can to take care of those children; it is my only way of helping the populations of the village hardly hit by this illness."

Apart from the 72 children registered at this improvised school dedicated to orphans, more than 60 other orphans from Zenie attended the community school which was built almost 10 years ago in very difficult conditions, without any support from the state or any other organization (per the statements of the community's representatives).

During exchanges with the teachers, one of the difficulties raised was managing the children's emotions especially when they mention, during their teaching, certain aspects that bring up the losses they have suffered. For instance, when they mention situations referring to the family, some of the children who have none left start crying. Being unable to deal with the emotional charge in these circumstances, teachers try to avoid referring to the family or to the parents during their lesson. One of the teachers who taught three classes at this school of about 100 students, stated: "One day, in the middle of a civic and moral education lesson, as soon as I mentioned the subject of the children's rights and the parents' obligations, many children started to cry because the subject reminded them of their deceased parents. This saddened me so much that I decided to select the kind of information to be taught to students."

10.2.4   BREAK UP OF THE SOCIAL COHESION: TENSIONS AND GUILT

If, for those reasons, some students do not wish to return to school to attend evening classes, others, however, do not wish to return home at the end of the morning, preferring to stay at school or with the teacher until the evening classes finish. The reasons identified by the teachers for those unusual behaviors are that some children are afraid of the "emptiness," or "absence" left by the death of their parents when they come back home or return to their foster families.

According to some people who were interviewed, there were many associations in the village before the crisis and the social ties were quite strong. The situation further reversed entirely to the outbreak. The community's solidarity groups which existed in the village had disintegrated and the mutual relations, which brought relief to people with difficulties, had deteriorated significantly, increasing the isolation of people at home and aggravating their distress. A woman in her thirties said: "there were many associations in our village but none has survived the crisis."

Moreover, the way in which the community managed the sanitation crisis gave way to a feeling of guilt. The people interviewed declared that they were against the intervention of humanitarian workers in their village at the beginning of the crisis, and they sometimes expressed their opposition violently, for instance, stone throwing. Therefore, they now think that they are responsible for the loss of their relatives. A man in his forties said: "We have been reluctant at the beginning and we were opposed, for a long time, to having the sanitation teams intervene in our village. This explains the heavy toll that we registered."

Furthermore, disagreements that created tensions among the villagers who faced the epidemic were mainly whether to comply with the traditions or follow the "humanitarian workers' recommendations," contributed with hindsight and with the analysis of the situation. Some blamed others for their attitude during the crisis which might have contributed to the deterioration of the situation.

One of the examples of this established fact is the eviction or resignation of most of the village authorities who were in charge during the crisis and their replacement with new authorities. The seven-position related to it were replaced and are now occupied by youngest (the oldest is 50 years old). Traditionally, the authorities are constituted of wise and much older village men. This reaction points out the fact that loss of trust has resulted from the authorities' members' management during the crisis.

Indeed, villagers think that the people who were in charge during the crisis did not help them apprehend the magnitude of the problem at the right time and did not take the right measures to break up the transmission chain. From the end of the outbreak till the day of their replacement in January 2016, the communication was broken between the old authorities and the citizens who always blamed the formers' lack of responsiveness in facing the crisis. Besides, the population thinks that this lack of responsiveness is one of the causes of the difficulties that they are facing today. Not only these authorities undertook any measure which could have helped them limit the damages during the crisis, but also remained idle when facing the consequences of this crisis, without requesting assistance and support to repair the damages incurred. On the opposite, the new authorities' initiative, to ask the psychosocial team for help to support the village has been notified: activities took place only two weeks after the first meeting.

Conversely, these old authorities rather accused the notability and the traditional healers who are the moral guarantors of the exercise for power in their village and even in most of the other villages, to have not been able to manage the crisis. It is believed that the epidemic entered the village via a witch doctor who took care of a patient without necessary precautions. But this version has created divergences inside the community, some doubting in the veracity of this hypothesis. Therefore, as per these old authorities, accusations toward them are generally perceived as unfair because they do consider to have always worked under the control of notables and witch-doctors who should be the first to take measures and not them. This research of responsible has contributed to create a climate of tensions and make more difficult the grieving inside the community.

10.2.5   IMPACT OF TRADITION AND RELIGION

The healers acknowledged that they have part of the responsibility as regard to the bad management of the situation and the extent of the crisis. Their analysis, posteriori, reflects that this was most probably due to their ignorance of the illness. During exchanges with the psychosocial team, they pointed out their needs of being better informed and trained, to be able to include recommendations in their practices and avoid similar sickness propagations. Their speech has been sometimes more radical. Indeed, some witch-doctors promised to not touch a sick person presenting a feverish condition without a check-up from a health center doctor attesting to the noncontagiousness of the patient.

All these attitudes, divergences pointed out, contributed to create division in the village. Those feelings were especially exacerbated by the low rate of recovery observed. In fact, only 7 healed against 118 deaths were reported, less than 5%, which is largely under lethality rate observed in the case of the epidemic. This high rate of mortality can be explained by the fact that patients went too late to the treatment centers and were already too weak to fight against the sickness, or they were not going to, at all, count on traditional healers to support them to deal with it. This assessment has even more deteriorated trust of the populations regarding those who took decisions, judged more damaging afterword. Healers and notability have seen their power drastically reduced.

Those divisions have also intensified, during a time, the gaps per religious affiliations. Before the epidemic, the Christian community of the village, in minority, was not widely accepted by the rest of the community because of differences and contradictions through religious practices and customs with other members of the village. During the outbreak, these tensions reached their height. Indeed, the Christians applied the measures of hygiene brought by the humanitarian workers and raised the animists' awareness to the necessity to apply them since there were risk behaviors. Therefore, animists accused Christians to connive with the humanitarian workers to spread the virus in their village. Even acts of violence occurred against Christians at some moments. The president of the Christian minority of the village claimed "What happened in this village was very hard. People thought that there were problems with witchcraft seeing the others die. They said at the beginning that seven people were to die, and then 20, but the people were dying in greater numbers. We have asked many times our non-Christian brothers to accept the humanitarian workers but to no avail. It is only at the end that they finally understood."

However, these tensions related to religious divergences subsided during the crisis as per certain testimonies collected, the Christian community also provided support and often visited the bereaved while everybody else was suspicious of the company of others. These acts of high humanitarian impact did not go unnoticed and they led the animists to have a different opinion of the Christians, which contributed to their reconciliation. This reconciliation is quite visible today, mainly within the members of the new bureau of the village authorities 4 out of 7 are Christians.

10.2.6   STIGMATIZATION OF ZENIE'S POPULATION

The scale of the epidemic within Zenie, as well as the attitude towards the humanitarian workers regarding the refusal of the villagers to apply prevention measures, led to the stigmatization and rejection of the people of Zenie by the neighboring villages. Certainly, the surrounding villages understood quickly that the village of Zenie was facing an epidemic and not a witchcraft problem. Besides, the decisions of the people of Zenie, when dealing with the situation, were strongly criticized and the refusal to accommodate the people of Zenie was quite clear on behalf of the other villages.

Most of the people interviewed in Zenie mentioned this feeling of rejection; many were thrown out of the villages where they tried to seek refuge. So, they lived isolated in their village for a long time. A woman in her forties recalls: "At that time when I left to find refuge in my village Lyezou, all of the inhabitants showed their disapproval and told me to leave their village." A man in his fifties said "During that period, no one from the surrounding villages would accept to come here and we were also forbidden from going to those villages ," and he added: "I felt sick one day in Irie and I collapsed when I was having my hair done. Everybody escaped when I told them that I come from Zenie. Only my son, who was with me, had the courage to help me and accompany me to our village, Zenie."

10.2.7   NUMEROUS AND LONG TERMS IMPACTS

These circumstances, heavy with facts and emotions, contributed to weaken resources which significantly aggravated the difficulties of the village. Lack of productivity related to the death of active members or depressed state of survivors, the combination of housework and income-generating activities or harvesting, the necessity of taking care of vulnerable people (and more specifically elderlies or orphans), the loss of social ties and solidarity among the villagers and the neighboring villages, have had a direct impact on the available resources in village.

Another consequence of the epidemic that is highly being felt today and which is spreading in the village is the hunger. This situation is experienced by most of the people and is particularly visible among children as per the teachers' statements.

For most of the students, only the school canteen which receives the support of the World Food Program (WFP) provides them one meal per day. Many children lack food at home and prefer to stay at school hoping to have one meal; they cling to their teacher. A teacher said, pointing to a student of less than 7 years old: "You see this kid, he doesn't want to go home; he wants to stay with me hoping to get something to eat. Unfortunately, I have no food today. I told him simply to wait in class. I became student parent by default and I am sometimes obliged to shave some students, to look after them. Besides, I would like to do more than that if I had the means."

The situation is also difficult for the teachers whose meals are not paid by the state and who no longer benefit from parental and other contributions as during normal circumstances.

These contradictions, reactions, tensions, and shortages may partly explain the crystallization of psychosocial problems noticed during the psychosocial procedures provided in the village.

10.3   THE FRC/MACENTA TEAM

10.3.1   COMPOSITION

The FRC/Macenta team includes 10 people: eight psychosocial counselors, one supervisor, and one expatriate manager. The psychosocial counselors work in pairs when dealing with the population hit by the epidemic. Working in a small team enables to facilitate expression and establish a circle of confidence. Moreover, the team endeavors to consider the gender element, creating a man/woman team as much as possible.

The psychosocial manager supervises the whole team that covers the organization of activities and the team management. He/she is mainly in charge of planning the work, training the teams, and ensuring emotional supervision.

10.3.2   TRAINING AND STRENGTHENING OF SKILLS

Since the start of the FRC operations at the end of 2014, a psychosocial team was trained for various activities aiming to improve their psychosocial skills, hence reinforcing the support to the populations suffering from psychosocial distress in general and, more specifically, in the framework of the Ebola crisis. Therefore, the provided training enabled enhancement or development of the following knowledge and skills:

10.3.2.1   THE SOCIAL APPROACH

Many training sessions have been proposed and carried out by psychosocial managers to enhance the teams' understanding of the aspects related to group dynamics, their possible impact on individuals and, conversely, the way in which individuals can influence these dynamics. Regarding the following concepts for instance:

"The group dynamics" based on the Karpman triangle or drama triangle, invented by the psychologist Stephen Karpman, is a method of representing the interactions that can exist among three individuals. Everyone plays a very specific role within these interactions: the role of a victim, a rescuer, or a persecutor. These roles influence the group dynamics and each player's own feelings at the same time.

KARPMAN'S DRAMA TRIANGLE

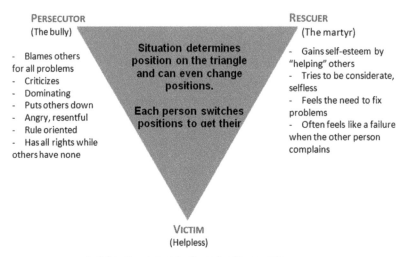

PERSECUTOR
(The bully)

- Blames others
for all problems
- Criticizes
- Dominating
- Puts others down
- Angry, resentful
- Rule oriented
- Has all rights while
others have none

**Situation determines
position on the triangle
and can even change
positions.**

**Each person switches
positions to get their**

RESCUER
(The martyr)

- Gains self-esteem by
"helping" others
- Tries to be considerate,
selfless
- Feels the need to fix
problems
- Often feels like a failure
when the other person
complains

VICTIM
(Helpless)

- Feels hopeless, trapped, ashamed, guilty, powerless
- Seeks others to solve problems, give them validation
- Refuses to make decisions, solve problem, or seek
professional help
- dependent

Adapted from Karpman's triangle: Karpman, S. (1968). Fairy tales and script drama analysis. Transactional Analysis Bulletin, 7(26), 39–43. https://www.karpmandramatriangle.com

"The triangle of abuse" is a result of the research work of Michel Dorais and Éric Verdier, and has also been addressed to identify the feelings and responsibilities involved in the context of discriminations. These authors highlighted the predominance of the scapegoat processes within violence phenomena against oneself and the other.

Taking into consideration the roles and emotions of the various protagonists enables to acquire efficient identification and analytical tools.

The psychosocial teams were acquainted with these aspects to better analyze and answer the discriminatory and exclusion attitudes which targeted the recovered and bereaved from Ebola.

10.3.2.2 PSYCHOLOGICAL APPROACH

Drawn from the manual "the community-based psychosocial support" of the International Federation of Red Cross and Red Crescent Societies

(IFRC) psychosocial reference center many aspects have been dealt with to strengthen the teams' understanding of the psychological aspects involved at the time of providing support to the Ebola-stricken people. These aspects should be considered to provide the appropriate assistance by understanding different attitudes observed in individuals. They mainly deal with the following modules and objectives:

- "Serious events and psychosocial support": Understand what a serious event might raise in terms of people's needs and define the psychosocial support and its role during these events;
- "Loss and bereavement": Define the whole set of circumstances surrounding the bereavement, describe the normal and complex bereavement processes and, finally, identify the best way to aid the bereaved;
- "The community-based psychosocial support": Understand how communities influence individuals and their psychosocial well-being, think of the importance of sociocultural context for each psychosocial activity;
- "Stress and adjustment" which aims to define stress, the frequent reactions that it may trigger, how to deal with them and the long-term consequences if not dealt with.

Finally, some specific training was provided, based on the psychoso-cial managers' specific skills, such as the depression and the suicidal crisis to define and identify them and to better deal with them when an individual is subject to them.

10.3.2.3   ACTIVITY MANAGEMENT TECHNIQUES

The provided training enabled the psychosocial counselors to adopt the appropriate attitudes during their sessions. This training was again based on the professional practice of psychosocial managers and "the community-based psychosocial support" manual of the IFRC psychosocial reference center. Therefore, modules were provided on: "the children's specific needs" which aims to ensure the children's security and well-being, identify the violence and ill-treatment inflicted on them, their reaction to stressful events, how do they cope and how to help them; "the manage-ment of a support group;" "the end of life care;" "the relation during the interview;" "the Carers-Cared for relation;" "the death announcement;" "the communication;" and so forth.

"The personal development: how can I assist others?" which contains modules on the relation of assistance, active listening, and reformulation was also used to train the team. This training was mainly based on the work of Carl Rogers, a humanistic psychologist, specialized in the assistance relation and who developed the principle of counseling. The method of Rogers is based on the person-centered approach, the empathy, and non-judgment.

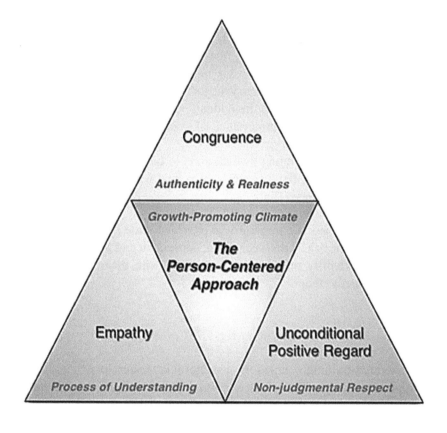

10.3.2.4 MANAGING THE COUNSELORS' STRESS

In addition to the support provided to the teams by the psychosocial manager, in terms of operational and emotional aspects through specific supervisions, training on stress and emotion management was organized. The main objectives were to identify the specific causes of stress felt by the humanitarian staff and volunteers (especially in the Ebola context), identify the overwork signs, each person's own resources which are available for the group as well such as support by peers, decompression activities (reading, sports, etc.) or the professionals whose help may be requested. These training sessions were based on the experience of psychosocial managers who provided them, as well as on manual of the participant of the IFRC, with the following module:

- "The suffering of carers"
- "Stress management and relaxation techniques"

10.4 THE IMPLEMENTATION OF THE RESPONSE TO THE POPULATIONS OF ZENIE

10.4.1 THE PSYCHOSOCIAL PROCEDURES CARRIED OUT IN ZENIE

In the framework of the procedures carried out in Zenie the activities proposed were:

Psychosocial talks/discussions: They aim to enhance reflection and raise awareness as to the social difficulties and the common needs, promote alternative solutions or new ideas to transform a problem or answer a certain need, and obtain acceptance and facilitation by the community leaders of the implementation of individual or collective psychosocial activities. The psychosocial discussions started at the beginning of January and involved 15 community leaders. In February, the team targeted all the population with the participation of 200 people and specific groups such as the wise men, the Christian community, and teachers. In March, these talks were proposed to the old and new authorities, the association of widowers, widows, and the student association.

Support groups: many groups were created to enhance the verbalization of real-life experiences and emotions, create or strengthen social ties and reduce the participants' psychosocial distress. The support group gathers generally 8–15 people who discuss a predefined subject to talk about conflicts and suffering and suggest possible means to resolve them. These discussions last about 45 min and represent an opportunity to speak without any interruption, judgment, and in strict confidence.

The support groups started to be implemented in this village since the first procedures were applied. In February, the groups were intended for the wise men, notables, men, young people, and women. In March, extra support groups were created for the traditional male and female healers.

These two activities (psychosocial talks and support groups) targeted specifically all the people of different age groups, various social classes, and gender, enabling every person's free expression on all crisis-related issues. Per the tradition of the village, people feel free to express their feelings and tackle sensitive issues if they were with their peers. These separate and small groups were also meant to avoid the tensions which were likely to emerge within mixed groups.

10.4.2   THE IMPACT OF FIRST ACTIVITIES CONDUCTED

The psychosocial activities proposed so far have had considerable positive impacts since the first procedures. They enabled reinforcing community resilience mainly by reducing the isolation of people through the activation of social ties, recreating speaking, and exchange spaces. The teams noticed that the people were more open to each other, some of them could resume work thanks to a better relationship with others and a renewed or enhanced self-confidence, as mentioned by a man in his fifties: "I want to thank your team, you are helping us a lot, I now manage to socialize with people and to go to work."

The work of the psychosocial team has also facilitated the creation of solidarity actions among groups of people facing the same problems. Therefore, the following groups were created:

- The group of widows: The group president stated: "We, the widows, created this group to help one another in the rural and all other activities, and we visit each other in the evening."
- The group of community support: One of its members explains the objectives and benefits: "The idea is to support and complete each other in our various activities, as your team mentioned the advantages. The men do the clearing and cutting down, then the women plow and weed, then we do the harvesting together. We are convinced today that this can help us immensely in dealing with the food problem!"
- The group of the young for the development of Zenie: As explained by one of the young members of the group: "We, the young people of Zenie, have been thinking after your first action (the psychosocial team's), to create a group to help our community, especially the orphans, further to this hard period of Ebola. We carry out solidarity actions such as the construction of sheds at a certain distance from the road to avoid traffic accidents and which could be used as orphanages. We also organize certain activities which benefit the elderly people."

The community approach enabled to identify the collective needs and support the villagers in the identification of their common resources. Further to the psychosocial activities, the people of Zenie started to plan seeking solutions to their problems, those which would often seem insurmountable until then. The community approach, mainly the support group activities, has also enabled to facilitate the verbal expression and to bring out individual needs. This assessment enabled to include, in the elaboration of the psychosocial support strategy of the people of Zenie, the organization of individual interviews to tackle specific problems in a more adapted way with those who have expressed the need. Therefore, an individual approach will be implemented to continue the support provided to the population and facilitate a global care.

10.5 LESSONS LEARNT

The Ebola virus has significant and immediate psychosocial impacts on affected populations, especially the bereaved or the people who recovered; impacts such as bereavement, anxiety, or exclusion. However, it is important to take into account the long-term effects resulting from the outbreak or the real-life experience on the social dynamics as well as individual suffering. The first psychosocial procedures are generally directed towards providing support to the sick people and bereaved in the management of bereavement or the reintegration of those who have recovered. However, it was noticed that the impacts continue and are transformed with time and they may increase the disintegration of communities and families: discrimination, rejection, and self-exclusion, difficulty in accepting the people recently taken care of by certain families due to the loss of those who used to meet their needs, and so forth.

In fact, the village of Zenie shows the importance of this vigilance initiated by the teams in the psychosocial approach carried out. They quickly provided self-expression spaces to the people of this village through the implementation of groups of expression and family and community mediations. These activities, which have been organized further to the community discussions, gave them meaning and an interest shared by all the members of the village, enabled to mobilize the population around common objectives, strengthen the solidarity, and reduce the loneliness of some people. The best understanding of the causes which generated such distress and tensions within this community enables to better support individuals in facing them, especially through verbal expression and the identification of individual and collective resources available to help them find sustainable solutions.

Finally, it is important to be careful in the emotional management of the staff when facing this crisis. Indeed, the life stories that they listen to convey emotional charges which may trigger a feeling of helplessness and/or remind them of a personal life story. That is why the psychosocial manager should ensure an operational, as well as emotional, supervision of his/her teams to help them manage their own emotions and enable them to act efficiently and sustainably.

The consequences and impacts of such a crisis last for a long time as being observed in the village of Zenie, and the psychosocial support must be pursued on the long-term to help the communities deal with their emotions, recover, and rebuild their lives. Therefore, it is crucial to ensure the continuity of providing activities and staff.

REFERENCES

Dorais, M. & Verdier, E. (2005). Petit manuel de Gayrilla à l'usage des jeunes: Ou comment luttercontrel'homophobie au quotidien Broché. Paris, France: F & O Publishers.

IFRC/PSP (2010). Psychosocial support based on the community. Copenhagen, Denmark: IFRC Psychosocial Support Centre.

IFRC/PSP (2012). Caring for the Volunteer. Copenhagen, Denmark: IFRC Psychosocial Support Centre.

(IFRC/PSP. (2015). Community-based psychosocial support. Copenhagen, Denmark: IFRC Psychosocial Support Centre.

Rogers, C. (1942). Counseling and psychotherapy. New York: Houghton Mifflin Company.

Rogers, C. R. (1977). On Personal Power: Inner Strength and Its Revolutionary Impact. New York: Delacorte Press.

Karpman, S. Fairy Tales and Script Drama Analysis. Transactional Anal. Bull. **1968**, 26(7)39–43.

Karpman, S. A Game Free Life: The definitive book on the Drama Triangle and Compassion Triangle by the originator and author-the new transactional analysis of intimacy, openness, and happiness, 2014.

CHAPTER 11

SAFE SCHOOL PROGRAM AS A PSYCHOSOCIAL SUPPORT TOOL IN ADDRESSING THREAD

JOSEPH O. PREWITT DIAZ

Chief Executive Officer, Center for Psychosocial Support Solutions, Alexandria, VA 22310, USA; jprewittdiaz@gmail.com

CONTENTS

11.1 INTRODUCTION

Threats to school are common and acknowledged by teachers, students, and parents alike. While conducting training of Safe Schools in 2003, the children were asked to conduct a mapping exercise of the school campus and report back the three most important threats. In Puri, Odisha, a group of children reported that on number one, in their list, were snakes in the classroom. In a similar exercise in Tanjabur, Tamil Nadu, some weeks later, the children responded with the main threat being the elephants breaking down the boundaries or walking into the school grounds during recess. The next question was "what do you do to reduce or eliminate that threat?" All the students in all grades in the schools knew how to address or reduce the threat. Human beings respond to stimulus based on reactions of the nervous system. The response will be retained and the individual will almost always respond to the threat with a series of learned behaviors that will reduce the threat. After conducting hundreds of these activities throughout India, the conclusion reached was that a "Safe School program" should be initiated with a mapping exercise naming the threats and possible solutions.

While we thought we had developed a good psychosocial component for schools, we were shaken up by two unpredictable disasters in schools in 2004. In Chaupali, Odisha, a lightning strike killed a child and left 37 others injured. Two months later a school fire in Kumbakonam, Tamil Nadu killed 99 children and left many others with significant burns in their body. The pre-training that some children had received in other parts of India would not have helped these children from the sudden occurrence of an event. Many months after these events we worked with parents, teachers, affected children, and communities to assist them to share their feelings (talk, recreate, participate in community activities, eat, and rest). We added the following to the Safe School activity. "If there is an event in school, get out of the building, gather in a predetermined spot, make sure that all your classmates are accounted for, provide first aid if anyone is hurt, and listen to your friends' stories, validate their feelings, and promote calm." Each student was responsible to assist the teacher until help arrived.

In resource-rich countries such as the United States, there have been a number of school shootings since the Columbine High School event in 1999, where 13 students were killed and 25 injured, that has motivated government agencies to come together and formally study the shooter behaviors (the US Secret Service and Department of Education, 2004), propose templates for plans (FEMA, 2013), and many training programs to teach teachers and other school personnel to respond to an active shooter event (such as ALICE, a 2-day training program that provides real-time suggestion for choices during a lock down). The reality is that all this guidance makes the reader aware.

This chapter will present three case studies and describe the steps of one psychosocial support program for schools called "Safe Schools." (1) Conduct a comprehensive assessment of the campus—building, grounds, and temporary structures, (2) identify threats and opportunities, (3) develop a plan based on prevention, response, and recovery, and (4) psychological first aid (PFA), counseling, and referrals for students and staff.

The remainder of the chapter presents: (1) brief section of the emotional and psychological impact of mass casualty incidents and (2) provision of psychosocial support for the affected community and management of mental health.

11.2   CASE STUDY 1

Two students from Columbine High School had been planning an act of destruction in school for some time. In April 1999, they entered the Columbine High School during the lunch break. They took some duffel bags and placed them in the cafeteria. They went back-and-forth to the parking lot. The reports indicate that they even told some students that they were going to do "something big" and suggested they leave the campus. The friends left and did not share the information with school authorities. As the shooting began, some students thought it was a school prank, "just some firecrackers." The students were shooting real guns and real bullets, others realized that a warning was given to students in the Cafeteria, but that was not sufficient as the shooters were already taking aim at some students. The shooters then went to the library where 56 students and four staff members had taken shelter. They had planned to explode homemade devices in the lunchroom with the objective of killing all students in that location (about 475). After roaming throughout the building, they committed suicide. After 47 minutes had elapsed there were 13 killed and 25 wounded and the two shooters had committed suicide Erickson (2001).

There were at least two factors that impacted the resolution of the threat: (1) lack of communication between the school and the incident command, and (2) personnel from multiple jurisdictions attempting to enter and comb the building. Some proactive behaviors saved lives during this event: (1) students ran away from the shooting and out of the building, (2) students and staff hid in safe spaces, (3) some in classrooms and other rooms locked doors and turned the lights off, (4) the students knew and trusted each other to find safety and comfort in the middle of chaos, and (4) students hid under tables, and other furniture, and kept quiet during the duration of the event.

As a result, the Secret Service and the Department of Education conducted a study in an attempt to understand the active shooter behavior before and during the event (Secret Service and Department of Education, 2004). The study yielded the following conclusions:

- Incidents of targeted violence at school are usually planned and well thought in advance. They have a scope and a sequence.
- Prior to most incidents, other people knew about the attacker's idea and/or plan to attack.
- Most attackers did not threaten their targets directly prior to advancing the attack.
- The profile of students who engaged in targeted school violence is not conclusive when similar events are compared.
- Most attackers are engaged in some behavior prior to the incident that caused others concern or indicated a need for help.
- Most attackers had difficulty coping with significant losses or personal failures. Moreover, many had considered or attempted suicide.
- Many attackers felt bullied, persecuted, or injured by others prior to the attack.
- Most attackers had access to and had used weapons prior to the attack.
- In many cases, other students were involved in some capacity.
- The event must be handled by the victims until law enforcement arrives. Most events take place in a 10-minute time frame. Despite prompt law enforcement responses, most shooting incidents were stopped by means other than law enforcement intervention. (the US Secret Service and Department of Education, 2004, pp 32–37).

11.3   CASE STUDY 2

On 16 July 2004, an alert was received at the Indian Red Cross Society Emergency Management Office in New Delhi indicating that a school in Kumbakonam, Tamil Nadu, India had caught fire and that 94 children were killed. The school had a total population of children of 692. This was the second largest in terms of casualties reported in schools in India.

A thatched kitchen near a classroom caught fire. Most of the dead children were in the two elementary classrooms near where the fire began. There were inappropriate fire exits that were closed so the children could not get out to safety. The school teachers were not trained to handle the situation and ran out of the building, many older students risked their lives to save the young children. The after action report commended that many of the older children for helping their younger brothers or sisters, broke walls and tin sheets to get out of the burning building, and provided first aid to the children that had managed to get out of the building but had some burns.

The after action report (IRCS Kumbakoman Fire Response, 2004) indicated that (1) the fire location of the open kitchen was too close to the classrooms, therefore the flames traveled fast to the classroom, (2) the teachers were not prepared to handle the evacuation of their pupils, they rather ran out of the building, (3) the exits were locked and children could not get out, and (4) the rapid actions of some older children and a janitor, who broke through an enclosure, facilitated the exit of the remaining children.

The assessment of the surviving parents and older siblings yielded some proactive behaviors that saved lives. (1) The older children were responsible for saving younger siblings and upon seen smoke ran to get their siblings and out of the main door. (2) A student reported that he had drawn a plan of the school and the grounds for a social study class the year before, and remembered some open spaces that had been covered with tin material, he took some children pushed his weight on the tin enclosure and they got out. (3) One of the children was a member of the Red Cross youth club in the school and had taken a first aid class.

What proactive actions saved the lives? (1) Familiarity with the campus both built and natural structures, (2) the custom that older children are responsible for their younger siblings, and (3) the quick action of a janitor and some older students that saved the lives of over 100 students.

In addition to the school being overcrowded, and the negligence of having an open fire to cook rather than a kitchen adjacent to a wooden classroom, there were some actions that may have prevented the tragic end of this fire. School staff and children were not aware of the perimeters of the school and the building within the enclosure. They were not aware that the smoke that they were seeing was a fire, in most days the smoke from the open fire was "a good thing," "the food is cooking we will soon eat" was the impression of one of the students interviewed. There was no plan for evacuation of the school.

11.4   CASE STUDY 3

On December 14, 2012, a lone gunman entered the Sandy Hook Elementary School, in Connecticut. After shooting away the front door lock entered the building, killing the principal and a school psychologist. He then walked down the hall past the main office and entered two classrooms where he shot four staff members and 20 children. The shooter then committed suicide. The attack lasted about 10 minutes and at the end, 20 children and 6 staff members had lost their lives but the impact of the attack will last a lifetime.

The result of this tragedy turned into some sorts of a blessing. A committee appointed by the Governor took a proactive view in exploring the action steps needed to guarantee the safety and security of children, school personnel, and the community. The first step was reframing the purpose of schools. The report frames that schools are neighborhood "hubs" Schmeiser (2015). Final report of the Sandy Hook Advisory Commission (Newtown, Conn: Sandy Hook Advisory Commission). The report was divided into three parts. One of those parts looked at the design and operations of Safe Schools from an all-hazard risk management approach (p. 4). The committee outlined and identified three components of the "Safe School." First, the design of the plan should be guided by the community, and strategies must be tailored to the specific needs of particular communities. Second, the standards to develop the plan should be nimble and updated as needs and environment changed. Third, the plan should have a representative group that would implement the Safe School plan (p. 5). The purpose of this approach was the assumption that a carefully planned and coordinated response would help to reestablish a sense of security, ensure needed services, and remain active as long as necessary. Ultimately, it would promote community-wide recovery (p. 9).

Schools are defined as "place" of care, and as such, it is important to understand that what happens in schools impacts all other "places" in the community (home and neighborhood). The report indicates the need to plan for healthy schools. Schools can be "healthy places" when the learning tools are geared toward positive development, and teachers and staff model healthy social development in place with the children-cognitive development (p. 99). This approach works best "when it is a component of school environment that informs the culture of the school and the behavior of adult educators" (p. 100). This type of educational approach fosters social problem-solving skills and helps in the development of a "sense of place."

In developing countries, the objective of the Safe School program is to teach children self-protection with support from their school community, social family, and community, through developing pro-social education. A major expectation is the learning of bonding with peers and teachers within the classroom and the greater school community that appears to be at the core of risk coping. Ultimately, the children develop psychological and social potential to face risk situations and recover with the support of their community. An example of a pattern to cope with risks is the ability to stay calm, or the ability to generate trust between the children and adults in the school.

11.5   PLANNING A SAFE SCHOOL PROGRAM USING PSYCHOSOCIAL SUPPORT STRATEGIES

A Safe School plan (Dayal, 2004) must be a multitiered intervention and support strategy that considers at least three important factors;(1) school-wide interventions focus on developing expected behaviors and psychosocial competence, (2) target student groups that are at greatest risk (linguistically diverse, disabled, and mobility impaired), (3) individual students included traumatized youth, whose cognitive or psychosocial needs are higher than the most students. Help them develop skills that support life and resilience during a crisis.

The previous section introduced three real-time examples of catastrophic events in schools. Two of those schools are in a resource-rich country and one school is in a resource-poor country. In resource, rich countries planning and response activities exclude the participation of children, although there is the inclusion of parents and community members. In resource-poor countries, children in all classrooms are part of the committees and have a responsibility for planning and caring for their peers (they become teacher assistants in times of emergencies).

This section introduces two different ways of preparing a plan for Safe Schools: (1) the developing countries (India, Indonesia, Maldives, and Sri Lanka) and (2) the developed countries (Sandy Hook, Newton, and Connecticut).

11.6   THE SCHOOL CRISIS RESPONSE PLAN IN DEVELOPING COUNTRIES: INDIA, INDONESIA, MALDIVES, AND SRI LANKA

A school response plan is a list of steps to be taken before, during, and after a catastrophic event to ensure the safety and psychosocial well-being of children and school personnel. The plan prepares the total school community to respond to a catastrophic event, to increase knowledge, and expertise in responding through practice. The assumption being that if people practice a response to a crisis situation, that on the day of the real event, they will respond automatically. The initial step in planning is gathering of all the concerned people (administrators, staff, and teachers) to discuss the importance of the planning and preparedness activities at the beginning of the year. The second step is to diagnose the threats and vulnerabilities on the school grounds. At this point, as part of the onboarding for the school year, all students walk around in their respective classes with their teachers and make notes about specific threats that they may see in the perimeter. It may be something as simple as animals running loose on the athletic field or electrical wires and connections uncovered, or evidence of mosquito breeding, and gray water seepage. At times students will also mention lack of a first aid kit, fire extinguisher, or locked exit doors after the school day starts.

The students return to their respective schools and prepare a map of the school premises that show the risks, resources, support institutions (police, fire, or emergency medical personnel), simply the custodian, or a teacher assistant. Once the risk and resources are identified each teacher and her classroom develop a response strategy that is based on a school intervention that proposes to "develop better-functioning schools and improve the physical and psychological health of students." To ensure this objective under the school intervention, the teacher prepares two lessons that expand over a semester time: (1) facilitating psychosocial well-being among the students through creative expressive activities, (2) enhancing health and hygiene knowledge and skills among the students.

The administration will request representation from students from each classroom, teachers, and support personnel to form a crisis-response team. The team will have five groups: (1) leadership, planning, (2) damage assessment, (3) evacuation, (4) physical first aid, and (5) PFA. The team coordinates, communicates, mobilizes materials as needed, monitors, and evaluates the results of the plan. The team is accountable to the parents, local authorities, and the community at large.

11.6.1 STEPS IN PREPARING A SCHOOL CRISIS RESPONSE MAP

11.6.1.1 PLAN

The planning team is responsible to execute the assessment and compiling results. It also has to compile the entire classroom plan and prepare a written document that acts as a guide to carry out activities of preparation, response, and mitigation of threats. It defines the problems, activities, and human and material resources that will be used.

11.6.1.2 EXECUTE

An administrator, a teacher, counselor or social worker maintenance personnel, school nurse, and two older children (boy and a girl) will compose the leadership team. The administrator is the single point of contact with external authorities. The counselor or social worker is responsible for teaching PFA to every classroom, the maintenance personnel are responsible for the buildings and grounds as well as helping develop an emergency evacuation plan, the nurse will be responsible for teaching first aid in every classroom. This group is responsible for the regular meeting of the teams, simulations, and drills.

11.6.1.3 EVALUATE

Identify the level of progress that has been achieved in the development of the response plan and propose modifications. Identify actions that support positive outcomes, and improve and strengthen the activities that do meet the objectives. Promote the participation of the school community in the improvement and execution of the plan.

11.6.1.4 CORRECT

The plan is nimble and is corrected as mistakes are found in the execution of the training, drills, and simulations. Each school in a district has its own plan; the activities vary in the perceived threats and the capabilities of the students and staff.

11.6.2 SAFE SCHOOL INTERACTIONS

The sessions with the students are also planned to ensure a child-friendly space, reinforce the value of a sense of the school as "place," and solution-focused activities to resolve challenges in the facility. Developing an awareness of the school grounds and understanding the school map, threats, opportunities, interactive games and play are practiced. The interactive games and play do ensure the faster recovery and gaining skills and knowledge among the children affected by disaster (play include fire brigade, hospital rescue, or lifeguard). The activities are also equally important for the students to enhance their psychosocial well-being in a general condition.

11.6.2.1 PLAY FACILITATES FOUR ASPECTS AMONG THE CHILDREN

The play is an important tool in recovery. Children can express themselves through drawing or reshaping plastic, and then share their products with others. The creative nature of the activity gives the child a sense of mastery. A detailed explanation of the effects of play is given in subsequent text.

11.6.2.2   GAIN MASTERY OVER EVENTS

Play helps children express feelings and emotions. This release helps children feel lighter. Repeated expressions also lighten the power of negative emotions it has over children's lives and help them move forward. Children can change their way of thinking, modify their behavior or learn new healthy ways of interacting, behaving or coping by reacting to concepts presented through stories, listening, and observing other children.

11.6.2.3   ENHANCE SELF-ESTEEM

Opportunities to present their creations, talk about themselves, get praised by others for their work or behavior, or make new friends help the children to feel good. While interacting and playing the children learn to take turns, share things, play by the rules, or to talk in a group.

11.6.2.4   MODELING OF BEHAVIOR

Preadolescents at the age of 10–14 years usually are highly influenced by the environment and also model the behavior that attracts them. The organized session on various psychosocial and health practices ensure modeling of pro-social behavior. At this age, preadolescents may experience greater peer influence. This influence could be used in a positive direction once the activities are taken up in a group and allowed to follow certain social behaviors that encourage group learning.

One of the behaviors that helped save the lives of young children in the Kumbakonam fire was that older students went to look for their younger siblings. The other behavior was that the students were aware of the topography of the campus and knew the ways of getting out of the campus without being burned or with a minor injury. Based on these examples the IRCS developed the school crisis teams as part of the Safe School program. These teams consist of teachers and students who are responsible for one task of immediate response during a crisis. Primary and secondary schools in India have limited resources in terms of money and facilities. When faced with a crisis event, these schools may not have the means of taking care of their immediate needs. They are entirely dependent on the relief and rescue teams and the community coming to their rescue and assistance. These teams help the school to respond to a crisis event and minimize the loss of life and property. These teams help students to move from being helpless victims to becoming energized, able victors.

11.7   PRINCIPLES OF SAFE SCHOOLS IN THE UNITED STATES

This section is based on the final report of the Sandy Hook Advisory Commission (2015, pp. 20–60). Most of the efforts in creating Safe School program in the United States are based on violent incidents such as "active shooter" incidents (i.e., Columbine High School and Virginia Tech incidents). Based on that history, the Sandy Hook report focused on developing a Safe School program with a focus on multiple risks as well as natural and man-made disasters. The Safe School plan should focus on site development, preparation, perimeter boundaries and access points, secondary perimeter, and the interior of the building (p. 21).

As in India, the perceptions in the United States is that schools are places of learning, cultural and social development, mastering of physical abilities, and should be the locus of community engagement (p. 24). Therefore, the school becomes an extension of values and virtues nurtured at home. In order for schools to do their job learning should take place in an environment free of fear, where students, teachers, and staff can interface in spaces that foster interpersonal interaction and efficacy.

The students and teachers have to learn about situational awareness (in the South Asia context this is called school mapping) that is a fundamental tool in behavioral observation, condition assessment, prioritization, and management of the incident. The context has to be such that teachers and students are able to observe, develop an awareness of what is going on around them and that at the first moment they see any suspicious behaviors they can react and share with school authorities. To reinforce this point, we were able to surmise that school shootings happen because they have been thought about and concocted in someone's head, are planned, based on prior information and environmental influences; the potential shooter has access to weapons and then proceeds to act (p. 26).

Schools should be places that are supportive and nonthreatening. Students and teachers should feel secure in an environment that enhances the roles of teachers and students. The school should facilitate the students' understanding of self-protection that reduces anxiety and promotes a sense of comfort, safety, and security. To achieve these feelings the school has to be nimble with characteristics that evolve over time.

The report suggested that a number of stakeholders participate in the planning of a Safe School physical and emotional environment; most of those are adults (parents, teachers, local decision-makers, public safety, and others). We agree in having participants such as school custodians and others that may have a greater awareness of the activities going on within and around the school. From the experiences in South Asia, one relationship that improved the Safe School program was the active participation of students in understanding the environment, planning proactive responses to threats. By involving the students, the voices of the "end users" were heard and the students felt responsible and accountable for the safety and security of their school.

School personnel should receive training, practical experience, and supervision in disaster-related behavioral health issues, including PFA, social support interventions, and bereavement counseling and support (Sandy Hook Report, p. 207). The training should be provided at the beginning of the year with follow-up meetings. Ideally, teachers and other staff members should be trained as a preparedness activity.

11.7.1   INTERVENTIONS

Disaster mental and behavioral health interventions are considered most effective when they are practical, flexible, empowering, compassionate, and respectful of the needs of affected individuals and their social systems (National Biodefense Science Board, 2008). Interventions should address both psychological and social/community needs. In addition, population monitoring, triage, and screening activities should be seamless, coordinated, and integrated to ensure that specific needs are addressed in a timely and appropriate manner. It is important to match actual need with the implementation strategy (e.g., who delivers the intervention and how the intervention unfolds) and the timing of the intervention.

Interventions must be developed and implemented to match different categories of disasters and the numbers of casualties (few, moderate, or mass casualty event) and reflect available resources over time (e.g., early dependency on locally available resources). This approach requires real-time disaster mental and behavioral (psychosocial) monitoring at a population/systems' level to identify emergent unmet needs and to inform a variety of tactical response and strategic planning. The design and implementation of an intervention must fit the ecology of the culture, place, and type of trauma at a systems level to optimize the effectiveness of any psychosocial treatment. Interventions aimed at facilitating adaptive coping in times of uncertainty attempt to achieve a sense of safety, calm, self- and communal efficacy, hope, social connectedness, ability to process complex health information, and other outcomes yet to be illuminated through research efforts (National Biodefense Science Board, 2008).

11.7.2   FACTORS THAT CONTRIBUTE AND EMOTIONAL RESPONSE

FEMA in 2004 prepared a pamphlet on factors that contribute to greater vulnerability. Here, we examine those factors based on the psychology of place (See Chapter 1 of this book). These include:

11.7.2.1   LOSS OF PLACE

Having been a direct victim of the destruction of the home, or being in the unfamiliar location and surrounding is associated with displacement including loss of daily routines and familiar space, toys, or odors. Being uprooted, seeing injured or dying people, being injured themselves, and feeling that their own lives are threatened.

11.7.2.2   HUMAN LOSSES

This includes the death or serious injury of a family member, close friend, or family pet and dislodging from their building or neighborhood.

11.7.2.3   TEMPORARY RELOCATION

Including temporarily living elsewhere, losing contact with friends, teachers, clergy, and neighbors, losing pictures or memorabilia that are important to the child, parental job loss, and the financial costs of reestablishing their previous living conditions.

11.7.2.4   PRIOR EXPOSURE TO TRAUMATIC EVENT

Displacement, migration or remigration, change of cultural or linguistic environment all these factors comprise prior exposure to the traumatic event.

How parents and caregivers react to and cope with a disaster or emergency situation can affect the way their children react.

The human response to an incident of mass casualty should be tailored to the phase, nature, and the background and experience of the responder. Children will react according to how their parents react and trusted adults provide safety and security, and create an environment of calmness. When parents and caregivers or other family members deal with the situation calmly and confidently, they are often the best source of support for their children.

10.8   KEY PERCEPTIONS OF SURVIVORS OF MASS CASUALTY EVENTS

Adult perceptions of loss, relocation, and reestablishment of place are directly related to the impact of a mass casualty event. The figure below introduces four key factors and reported consequences: (1) subjective experience, (2) psychological impact, (3) worldview, and (4) stigmatization (Fig. 11.1).

Key Perceptions of survivors of mass casualty events

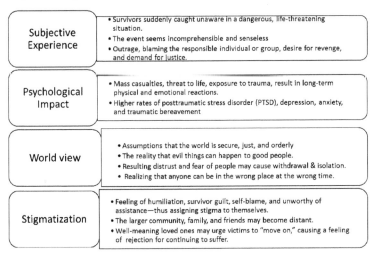

FIGURE 11.1   Key perceptions of the survivors of mass-casualty events.

Subjective experiences should go with the meaning of an event to a survivor. Norris (2007) suggests that the psychological consequences of directly experiencing or witnessing a mass shooting are often serious. It may seem that the event raised feelings of unbelief, or that the survivor was caught off guard, and lacked the coping skills to address the consequences of the event. There is a fear of losing their lives and may seem impossible to cope with the immediate results. The evidence of emotional responses of primary victims suggests the need for psychosocial support.

Norris (2007) reported that at less severe levels of exposure, the impact of mass shootings extends far beyond the primary victims to encompass the community. Community members resent the media intrusion, the sense that they are being blamed for violence and the convergence of outsiders. The reluctance of some members to focus on the event, while others need to, is consistent with community dynamics observed after other types of disasters (p. 3).

Witnessing a school shooting or surviving a school fire where many of your friends have died, can have emotional, psychological, and physical effects. These effects include nightmares, resisting the return to school, regressive behaviors, headaches, stomach problems, and sleeping problems. Other symptoms related with loss of place can impact children in several ways such as a change in school performance, changes in relationships with friends and teachers, anxiety, and loss of interest in activities that were once enjoyed.

Children are not the only ones affected by school shootings. School personnel also can develop psychological problems. Teachers who have witnessed a school shooting suffer from effects like not feeling safe, withdrawal, divorce, absenteeism, and feeling a lack of support. In the Kumbakonam, fire teachers blamed themselves for the fire and death of children and felt that the community blamed them. Many moved to other communities or have quit teaching.

Worldview is best explained by making meaning of the disaster. The psychosocial impact of a mass-casualty event is cultural and contextual. It has to do with age and life experience, and knowledge about what social forces that are impacting the survivors in the immediate aftermath, recovery, and long term. Worldview will lead us to help the neighbors and to mobilize the community to attend to the needs of the survivors.

The contexts dictate the reactions of the survivors. For those living in poor communities where school fire is just another event in the communities' everyday life, may not be impacted other communities. The reality is that for affected families' struggle over a period of time to make sense of the disaster and its impact in the families; they will rekindle the desire to move forward and plan to reestablish the family in a new place. Walsh (2007) reports that "certain core beliefs ground and orient people, providing a sense of reality, meaning, and purpose of life. Assumptions may be that others can be trusted, communities are safe, the future is predictable, and children will outlive their elders (p. 211)."

The stigma created by a lack of experience and training on how to handle a school event generates a feeling of stigma on teachers and other school personnel. They also have the effect of self-guilt and a feeling of responsibility for the bad things that happened. Stigmatization places teachers and school personnel to (1) question their belief system, (2) distorted organizational patterns such as stigmatized loss, and (3) problem-solving capacity is overwhelmed and becomes disorganized. Survivors face a hard and lonely road to recovery. In the Kumbakonam fire two girls survived after jumping out of windows. The initial public events acknowledged these girls as heroines. However, weeks after the fire the mother of one of the girls reported that she was withdrawn and did not want to go to school. The girl reported that no one had asked her about her experiences, and she did not want to reveal her shame of having jumped over her peers to jump out of the window and save herself. The other child had experienced a back injury, initially welcomed home later refused to be seen in public because of the scars on her face and arms.

11.9 PREVALENCE OF PSYCHOLOGICAL AND SOCIAL ISSUES

Lowe and Galea (2015) report three predictors of adverse mental health outcomes: (1) demographics and pre-incident characteristics, (2) exposure to the incident, and (3) post-incident psychosocial resources.

Children, elderly, a population with special needs, and those with preexisting mental health needs are most in need of psychosocial support in the aftermath of a mass-casualty event. The figure below presents the impact of the event and approximately time of recovery as suggested by prevalence studies (Fig. 11.2).

Prevalence of Psychological and Social issues over time

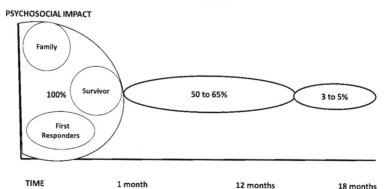

FIGURE 11.2   Prevalence of psychological and social issues over time. (*Source:* This figure is based on Lowe and Galea (2015)).

The event of mass casualty affects the community; the primary survivors are probably the most affected in the cone of probability. First responders are affected the most, often time they get injured trying to save lives and property. The families of survivors are affected, however, they will feel better with timely and accurate information and proximity to their dear ones. The community is impacted by proximity to the location of mass casualty, and by news articles and television scenes. The inability to take action to make dear ones feel better and the community to heal tends to increase until the survivors begin to get discharged from hospitals, and memorial events provide an opportunity for the community to vent, pray, and talk about their feelings and the time frame for this initial response is between 48 and 72 hours.

As time goes by between 1 week and 18 months, about 50–65% of the population will benefit from community activities, memorials, and other religious and social activities. Many of these activities are psychosocial spaces where the general population and those directly affected will be able to express their feelings and accept what has taken place, develop some coping skills to move on with their lives, and become more resilient.

About 3% of the survivors will continue to need additional community, and/or other specialized medical services. Most of these services will be a result of preexisting psychological or psychiatric issues, combined with complex physical trauma.

Evidence-informed assumptions (Lowe and Galea, 2015; Hobfoll et al., 2011) suggest a broad range of psychosocial responses in which fear and anxiety are the most common. Threat tends to exacerbate emotional and psychosocial responses that manifest themselves in an increase of health-care consultations.

There are some positive reactions that foster recoveries such as social support, receipt of individual counseling or PFA, and avoidance of the affected buildings. In most mass-casualty events, the immediate response is for neighbors to help neighbors (the Good Samaritan effect). Protective factors contribute to longer-term outcomes. Such protective factors include the availability of someone to talk to, personality characteristics, beliefs and attitudes, coping styles, and social relationships (Stephenson et al., 2009).

11.10   PSYCHOLOGICAL FIRST AID IN SCHOOLS

PFA is an evidenced-informed tool devised in a modular approach to help children, adolescents, adults, and families in the immediate aftermath of a mass-casualty event (Ubernik and Husson, 2009). The tool is designed to reduce the initial distress caused by traumatic events and to foster short- and long-term adaptive functioning and coping. PFA best serves the affected people if considerations about linguistic, cultural, and contextual are taken into account. It best serves the receiver when the helpers are representative of the affected community.

PFA is based on an understanding that disaster survivors will experience a broad range of early reactions. Some of these reactions will cause enough distress to interfere with adaptive coping, and that recovery may be helped by support from compassionate and caring responders (NCTSN, 2006, p. 5).

An important aspect of recovery is to treat the emotional side effects of violence and stress. PFA is an evidence-informed, modular approach used by disaster response workers to help individuals of all ages in the immediate aftermath of disaster and terrorism. PFA is designed to reduce the initial distress caused by traumatic events and to foster short- and long-term adaptive functioning and coping (DHS, 2015).

PFA does not assume that all survivors will develop mental health problems or long-term difficulties in recovery. Instead, it is based on an understanding that disaster survivors and others affected by such events will experience a broad range of early reactions (e.g., physical, psychological, behavioral, and spiritual). Some of these reactions may cause enough distress to interfere with adaptive coping, and recovery may be helped by support from compassionate and caring disaster responders (DHS, 2015).

PFA is a tool for early information gathering to help providers make a rapid assessment of the students' immediate concerns and needs, and to implement supportive activities in a flexible manner. It relies on field-tested, evidence-informed strategies that can be provided in a variety of settings. The intervention emphasizes developmentally and culturally appropriate interventions for persons of various age groups and cultural and linguistic backgrounds. It includes handouts that provide information for all age groups for their consideration over the course of recovery (NCTSN, 2006, p. 6).

Brimmer et al. (2012) have prepared the 2nd edition of PFA for Schools. They offered a five-point explanation of the importance of psychological support for the schools that is briefly mentioned below. Schools are typically the first service agencies to resume operations after a disaster/emergency and can become a primary source of community support during and after the incident. Schools are where children spend a majority of their day and where they receive substantial support from teachers and other staff members, and some schools are the primary setting for psychosocial support and child mental health services.

In many ways, teachers and staff are the "first and last responders" for children in an emergency. Students look to their teachers and to school administrators for leadership and guidance, while parents expect and demand that school personnel respond competently and appropriately in such situations. In a school-wide emergency, children's "everyday" school personnel can provide much of the intervention needed to stabilize the situation.

Preparing for emergencies is critical for all school staff. While school personnel should be prepared to respond to high impact/low-frequency events such as school shootings, large-scale natural disasters, and public health emergencies, they must also be prepared to address smaller scale events that schools face each day. Emergency events that do not typically garner national headlines, but do disrupt the learning environment, such as suicides, transportation accidents, peer victimization, community violence, staff or student deaths, injuries on the playground, and infectious diseases outbreak.

Preparedness involves (1) having a comprehensive response and recovery plan, (2) training staff to address the immediate, midterm, and long-term needs of students and staff members, (3) frequent practicing of the comprehensive response and recovery plan, and (4) evaluation and redesigning of plan components that no longer meet operational standards for the school. When people are trained in emergency protocols (including students, in the middle and high school) and have knowledge of techniques to reduce anxiety and establish calm, they are better able to handle the emergency and be of help to the people affected.

Emergencies affect students' academic and social achievements. Having an effective school psychosocial and mental health-recovery plan in place, which includes interventions such as PFA-S; it is critical when emergencies threaten to significantly disrupt the learning environment. Such events adversely affect students' academic and social performance frequently. Counseling services and programs addressing students' developmental needs have traditionally been viewed as supplementary services, "add-ons" to the academic mission of the school. However, when students' psychosocial and mental health needs are addressed in a developmental, systematic, and comprehensive manner, students achieve a higher level.

Trauma-related distress can have a long-term impact if left untreated. Unaddressed mental health needs, including those from exposure to violence and other potentially traumatic events, increase dropout rates, lower academic achievement, disrupt peer relationships, and impact overall well-being. Thus, PFA is important for the well-being of children and school personnel, but are also critical for the central educational mission of schools.

Brief interventions can produce positive results that last. A growing body of research shows that there are brief, effective interventions that have a long-lasting positive influence on students' and staff members' trauma-related distress. PFA-S draws from the best available evidence identifying factors that promote improved student and staff functioning after disasters and other emergencies (Brymer et al., 2012).

Response: Their reaction may be different. Provide PFA at the site for teachers and students refer to MH setting if needed. The action is taken to effectively contain and resolve an emergency and to decrease the potential for such an emergency to escalate. During this phase, the school executes the emergency-management plan and emergency procedures and initiates preliminary activation of the PFA-S teams. Although the response phase may have a clear ending point for emergency-response agencies, the transition into the fourth phase, recovery, may be less distinct. Prepare clear and precise communication for parents and dear ones that may be waiting.

Recovery: In the aftermath, all the affected people will be impacted. Steps should be taken to assist students, staff, and their families in the recovery process and to restore educational operations in schools. The very early stages of the recovery phase (hours or days after an emergency) are the most appropriate time to deliver PFA-S. Depending on the nature of the incident, recovery may be a long-term process. PFA-S is an acute intervention.

11.11 PROVIDING PSYCHOSOCIAL SUPPORT

After a mass-casualty event, mental health and psychosocial support activities are initiated immediately. The resources are deployed to several locations (on scene, community vigils, memorials, and affected schools and community organizations), and to physical facilities (hospitals, family assistance centers (FAC), first responders' organizations—clubhouse). Psychosocial support is found within the community of volunteers and paid staff and survivors that are transitioning from the hospital to their home.

Demographic and health survey (DHS, 2015) has developed guidance that addresses the role of a family reunification center and the specific psychosocial needs of the affected families. Where the immediate reunification of loved ones is not possible, providing family members with timely, accurate, and relevant information is paramount. The local or regional mass fatality plan may call for the establishment of a FAC to help family members locate their loved ones and determine whether or not they are among the casualties (American Red Cross, 2017). This center should be placed away from media view or exposure and it is recommended that the families of the victims be separated from the family of the active shooter. Although the FAC should be away from the incident command, care should be taken to ensure that it is not so far away from the incident site that family members feel excluded.

DHS (2015) suggests essential steps to help establish trust and provide family members with a sense of control. This can be accomplished by identifying a safe location separate from distractions and/or media and the general public, but close enough to allow family members to feel connected in proximity to their children/loved ones; scheduling periodic updates even if no additional information is available; being prepared to speak with family members about what to expect when reunited with their loved ones; and ensuring effective communication with those who have language barriers or need other accommodations, such as sign language interpreters for deaf or hard of hearing family members. When reunification is not possible because an individual is missing, injured, or killed, how and when this information is provided to families is critical. Before an emergency, the planning team must determine how, when, and by whom the loved ones will be informed if their loved one is missing or has been injured or killed, keeping in mind that law enforcement typically takes the lead on death notifications related to criminal activity. This will ensure that families and loved ones receive accurate and timely information in a compassionate way.

Victim and family support is a critical component to ensure a successful overall response to a critical incident. It is important to make sure that the response is coordinated through each phase including the immediate response, transition process, and post-crisis support in a way that integrates into the investigative and operational response. There are predictable challenges and practical solutions in mass-casualty events. Coordination with local resources is critical to ensure a smooth provision of services throughout the longevity of the case. The quality of the overall operational response to a mass casualty, in large part, will be judged by the response to victims and families and should be based upon trust, cooperation, and respect shown to victims, families, and eyewitnesses. Response planning should always track and adjust to meet the needs of the victim/family and the dynamics of the situation (DHS, 2015).

11.11.1   CRISIS COMMUNICATION

Communication during an incident is critical. Whenever possible, communication should be carried out in plain language. If needed, communication should be conducted in several languages. There should be a standardized communications plan to ensure all responding partners can communicate. This should include establishing the use of common terms to describe actions, locations, and roles. Communication plans should also include early notification of the health care system and facilities that may be called upon to receive casualties (DHS, 2015).

Communication is an essential mental health and psychosocial support intervention in disaster preparedness and response. Information during a crisis can facilitate lifesaving and altruistic behaviors and discourage actions that impede an effective response or rend the social fabric. Disaster mental health and psychosocial support specialists with expertise in crisis and risk communication can play important roles in maximizing effectiveness in communicating with the public during crises (National Biodefense Science Board, 2008).

Disaster-related communication efforts have focused heavily on the development of key messages presented, ideally, in "one voice" (i.e., one-way communications from authorities to the public). While this approach entails understanding and anticipating the public's information needs, it captures neither the richness of the public's ability to help solve problems nor the complexity of information needs that prompt people to take protective actions. Moreover, it does not fully take advantage of what has been learned about the ways people behave in disasters. For example, evacuation behavior illustrates the importance of informal networks and the role of nonverbal communication. Consider that people generally decide whether to comply with an official warning to evacuate only after gathering more information to validate the threat. This additional information comes through consulting with family, neighbors, friends, or coworkers and by checking other news sources. Establishing a true dialogue and engaging with the public before, during, and after an event promotes a sense of collective competency in mastering the challenges posed by the event and a sense of collective efficacy, essential ingredients to a community's resilience (National Biodefense Science Board, 2008).

The identification of groups at higher risk for psychological distress can inform the development of special messages targeted for these populations. Many disasters affect "hidden victims," that is, those who do not readily come to mind as being under stress, yet. For example, building inspectors, who must tell families that they cannot return home to damaged buildings, can be under tremendous stress. Developing news reports or using other means to educate the public about why buildings are not habitable, and acknowledging the difficulty of not being able to return home may help members of the community focus their anger on the disaster rather than on a government official (National Biodefense Science Board, 2008).

11.11.2   FAMILY ASSISTANCE CENTERS (FACs)

In the post-event stage, FEMA and voluntary agencies such as the Red Cross staff and volunteers provide assistance to survivors by staffing FAC is the next link in the modification. This is a spot for linkage between survivors, local and federal agencies, and families. In the immediate aftermath of a mass-casualty event such as the Orlando shooting (Summer 2016), family and friends were looking for information about their friends or family. In Orlando, there were reports of the use of social media as a source for finding out information about dear ones. When that resource did not work, people were calling or physically showing up at hospital emergency rooms. This is a natural reaction, family and friends will gravitate to where they believe their dear one might be or where they think they can find information.

The FAC is important because this facility can provide the information that is sought by family and friends. This facility provides other functions as well: (1) synchronizes mental health and psychosocial services for all that may identify as family members (sometimes up to ten people), (2) facilitates communications with other government and nongovernment agencies, and (3) provides support to responders and minimizing the impact of the event, (4) makes the necessary arrangements, and supports memorial services, visit from public figures from outside the impacted area, and for transportation of family and friends traveling to and from the impacted area.

The Red Cross has a specialized team of mental health workers that work in FACs. This planning team encourages and empowers local government and community organizations to take charge of the response. The role of the team is to listen and connect, assist the survivor to put closure to the emotions generated by the event and to attend to high-level visits.

11.12   LESSONS LEARNED

There are some lessons learned from participation in post-event activities.

1. In the immediate aftermath of a mass-casualty event, children, special needs population, and the elderly will benefit from PFA.
2. Encourage local involvement in all phases of the recovery effort.

3. Calm reassurance can help family members feel safer, and more in control of their lives.
4. Plan to serve the whole community. The exposure of mass-casualty events is far-reaching.
5. Disaster mental health volunteers must check themselves and remain in their place. Exhibit compassionate presence, practice self-care, and exhibit self-care by practicing stress management activities.

11.13   SUMMARY

This chapter presented three case studies of recent mass-casualty events. It elaborated on the reactions of the survivors, their relatives, and the community by presenting research on the impact of mass casualty events on survivors, their family, and the community. Four key feelings of the impacted population were presented and discussed. The tools and systems for response were explained, and the Federal Regulations involving the Red Cross were mentioned. Five lessons learned were collected.

REFERENCES

American Red Cross. *Terrorism—Preparing for the Unexpected*. American Red Cross: Washington, D.C., 2001.
American Red Cross (2017). Disaster Mental Health standards and procedures. Washington, D.C.: American Red Cross Disaster Cycle Services.
Biodefense Science Board. *Disaster Mental Health Recommendations*. National Biodefense Science Board: Washington, D.C., 2008.
Brymer M.; Taylor M.; Escudero P.; Jacobs A.; Kronenberg M.; Macy R.; Mock L.; Payne L.; Pynoos R.; Vogel J. *Psychological First Aid for Schools: Field Operations Guide*, 2nd ed.; National Child Traumatic Stress Network: Los Angeles, 2012.
Dayal, A. *The Kumbakonam School Fire Tragedy: Final Report*; International Federeration of the Red Cross: New Delhi, India, 2004.
DHS. *Planning and Response to an Active Shooter: An Interagency Security Committee Policy and Best Practice Guide*; Department of Homeland Security: Washington, D.C., 2015.
Erickson, W. M. *The Report of the Columbine Review Commission*; Governor's Office: Denver, Colorado, 2001.
FEMA. *Helping Children Cope with Disasters*; FEMA and American Red Cross: Washington, D.C., 2004.
Hobfoll, S. E.; Canetti, D.; Hall, B. J.; Brom, D.; Palmieri, P. A.; Johnson, R. J.; Galea, S. Are Community Studies of Psychological Trauma's Impact Accurate? A Study among

Jews and Palestinians. *Psychol. Assess.* **2011,** *23*(3), 599–605. http://doi.org/10.1037/a0022817.

Lowe, S. R.; Galea, S. The mental health consequences of mass shootings. *Trauma Violence and Abuse.* Research Gate. Downloaded from tva.sagepub.com at Columbia University, 2015 (accessed June 18, 2015).

NCTSN. *Psychological First Aid,* 2nd ed; NCTSN: Washington, DC, 2006.

Norris, F. H. Impact of mass shootings on survivors, families, and communities. *PTSD Res. Q.* 2007, *18*(3), 1–3.

Office for Victims of Crime. *Responding to Victims of Terrorism and Mass Violence Crimes;* OVC Resource Center: Rockville, MD, 2012.

Orcutt, H. K.; Miron, L. R.; Seligowski, A.V. Impact of Mass Shootings on Individual Adjustment. *PTSD Res. Quaterly,* **2014,** *25*(3), 1–3.

Prewitt Diaz, J. O.; Bordoloi, S.; Mishra, S.; Rajesh, A. *School Based Psychosocial Support Training for Teachers;* Indian Red Cross Society: New Delhi, India, 2003.

Stephenson, K.L., Valentitiner, D.P., Kumpula, M.J. & Orcutt, H.L. (2009). Anxiety Sensitivity and Posttrauma Stress Symptoms in Female Undergraduates Following a Campus Shooting. J Trauma Stress. 2009 December ; 22(6): 489–496. doi:10.1002/jts.20457.

U.S. Department of Education, Office of Elementary and Secondary Education, Office of Safe and Healthy Students, Guide for Developing High-Quality School Emergency Operations Plans, Washington, DC, 2013.

Walsh, F. Traumatic Loss and Major Disasters: Strengthening Family and Community Resilience. *Fam. Process,* **2007,** *46*(2), 207–227.

Weaver, J. *Disasters: Mental Health Interventions;* Professional: Bethlehem, PA, 1995.

CHAPTER 12

PSYCHOSOCIAL SUPPORT AND EPIDEMIC CONTROL INTERFACE: A CASE STUDY

JOSEPH O. PREWITT DIAZ

Chief Executive Officer, Center for Psychosocial Suport Solutions, Alexandria, VA 22310, USA

CONTENTS

12.1 INTRODUCTION

An epidemic is the spread of a contagious disease within a specific geographical area or population. Epidemics are often characterized by diseases that can spread rapidly, especially in poor and underserved communities, that lack of access to water or basic hygiene, or where water accumulations (potholes) form, or open drains. Dengue, chikungunya, or Zika are transmitted through the mosquito *Aedes aegypti*, whose hatching places are near the homes. Once the mosquito bites several people and they contract the virus. The transmission occurs from human to humans.

The risk of an epidemic increases during the rainy season, where water is insufficient and insecure and inadequate sanitation can put individuals at greater risk of contracting a contagious disease. In the case of the Zika epidemic, most mothers with children with microcephaly report high fever, rash, conjunctivitis, and/or influenza.

People affected by epidemics may experience high levels of stress, which in extreme situations can be debilitating. High levels of stress are often prevalent in communities where knowledge of the disease, how it spreads is limited and where the risk of transmission is high. Not only those who have fallen ill, but also those affected or are associated with the disease, such as members of the family of an infected child with a child, health workers, or people who have suffered complications as a result of the Zika epidemic, may be vulnerable to social stigma. This, in turn, can bring economic consequences and other losses that significantly disrupt a person's daily routines and sense of normalcy In the same way, mothers can experience anxiety, sadness, anguish, guilt, concentration problems, and fear. In such situations, psychosocial support may be useful or necessary to help individuals recover.

Psychosocial support helps people recover from a critical situation that has changed their lives. One intervention may be community-based psychosocial support interventions. This intervention focuses on strengthening the social ties of affected people in communities by improving the psychosocial well-being of individual's neighborhoods (IFRC Psychosocial Framework 2005–2007).

This approach is based on the idea that if people are empowered to take care of themselves and each other, their individual and community trust will have improved as well as their resources. This, in turn, will encourage a positive recovery and strengthen your ability to deal with the challenges of the future. Psychosocial support can be both preventive and rehabilitative. It is preventive when it decreases the risk of developing mental health problems. It is a rehabilitator when it helps individuals and communities overcome and deal with psychosocial problems that may have arisen because of shock and crisis effects. These two aspects of psychosocial support contribute to the building of resilience in the face of new crises or other challenging life circumstances.

12.2   THE INTEGRATION OF PSYCHOSOCIAL SUPPORT INTO THE ZIKA PREVENTION CAMPAIGN

This section will focus on sharing experiences about how psychosocial support interventions were integrated into the Zika prevention campaigns in three countries: Puerto Rico, Dominican Republic, and Brazil The chapter is divided into four parts: context, a summary of experiences in countries visited, conclusions, and recommendations.

12.2.1 CONTEXT

12.2.1.1 BRIEF DESCRIPTION OF THE BEGINNING, EVOLUTION, AND CURRENT SITUATION OF THE ZIKA VIRUS IN THE REGION

Already by 2013, an epidemiological study reported the presence of the Zika virus in Americas (http://science.sciencemag.org/content/351/6280/1377. full). The Zika virus has been detected in Americas since February 2014. Chile's public health authorities confirmed the first case of Zika virus infection on Easter Island (Chile). The sequalae of the virus include fever, rashes on the skin, and headache. It was found that pregnant women who contracted Zika could affect their children with microcephaly or a neurological disease called Guillain–Barré syndrome. Already by February 1, 2016, the Zika was declared as a global epidemic and the World Health Organization (WHO) as an emergency.

In 2015, an outbreak of acute febrile illness was reported, including fever, skin rash, myalgia, arthralgia, and headaches. After laboratory tests, none of the samples were found to be positive for dengue, chikungunya, rubella, or measles. A few weeks later eight cases were reported in the State of Bahia that was positive for the Zika virus. It is not until February 17, 2016, that the Ministry of Health publicly reported the Zika virus as a disease (PAHO, 2016).

As there are no means to prevent the virus through inoculation, and the potential number of people affected (between 4 and 5 million), the Red Cross and Red Crescent requested and obtained funds to set up a prevention program in five affected countries in Central and South America. The American Red Cross supported prevention and psychoeducation programs in Puerto Rico and the Dominican Republic/Haiti in 2016 through its domestic and international programs. The domestic program in Puerto Rico included psychoeducation for mother's in government hospitals and youth activities in the public schools. The Red Cross International program in Haiti developed materials and included a psychosocial support that was identified as a cross-cutting theme to be addressed in the eradication campaign.

12.2.1.2   EPIDEMICS HAVE AN IMPACT ON MENTAL HEALTH AND PSYCHOSOCIAL PROCESSES IN THE SHORT, MEDIUM, AND LONG TERM

Epidemics can be characterized as low-intensity, but long-lasting events. In order to understand the impact of this emergency on addressing psycho-social aspects of the population within the framework of the Humanitarian Charter and Minimum Standards for humanitarian response, because it is important to ensure that physical and psychosocial well-being goes hand in hand. In the case of the Zika epidemic, the short-term psychosocial impact lasts between 3 and 14 days.

The impact of psychosocial support is limited to accompanying, providing safety, and calming the patient. In the medium term we consider pregnant mothers, and the families they support, this intervention lasts approximately nine months and requires psychosocial support and follow-up, as well as psychoeducation about the possible causes of the virus in the unborn child. Long-term interventions will be given to children with microcephaly and Guillain–Barré syndrome. This intervention requires the participation of government agencies alike, the community and the Red Cross. The goal is to introduce and accept children with microcephaly as other children in the community.

The following are five possible sequels of the impact of microcephaly and Guillain–Barré syndrome and affected families: feelings of fear and impotence.

12.2.1.2.1   Fear and Impotence

Mothers and their families were struck by fear and helplessness at the unpredictable impact of the Zika epidemic. The members of the communities visited countries which found it difficult to overcome the fear and the uncertainty of the results of the Zika virus during the pregnancy, and more are affected those mothers whose children were born with microcephaly. A young mother of about 21 said, trying to suppress her tears:

> "my world dissolved when I learned that the child would be born with microcephaly. Hope left the house, and with a deep sense of guilt, my mother with tears in her eyes, reminded me that I would have to change my lifestyle once the baby was born. I could no longer finish my studies

and I would have to dedicate myself to taking care of my daughter all the time."

The grandmother of a child with microcephaly told us:

"when the baby is born, our lives will change, we will be alone, my daughter, my granddaughter, and me. Nothing will be the same anymore, our life as we know it will no longer exist." "The community feels help-less by the effects of the damn mosquito."

The family lost the desire to continue living a normal life in their neighborhood Although, some of the neighbors of this family continued to support them.

Mothers and relatives interviewed mentioned a lack of "desire or energy" to undertake or resume studies and/or economic activities. My daughter sits on the balcony and cries since the creature arrived. A mother of 16 told us that "sometimes when she goes to a warrant, she begins to cry and returns directly to my house. I feel no encouragement at all."

The Zika epidemic has weakened the desire to undertake and build future projects, affected communities are unable to overcome the gap left by affected children and the deceased ("my grandfather died as a result of Zika, I loved him very much and made him less every day") and are reluctant to accept the daily reality that for some mothers has become increasingly difficult.

The blow suffered by the families is enormous and most of the family, close to a girl with microcephaly of eight months of age, expressed discouragement, or a persistent feeling of impotence. Neighbors and rela-tives also commented that this situation and the outcome in children with microcephaly, change community networks to other children (who are not affected) feel as a trauma that affects their psychological well-being. For the most vulnerable people, the trauma created a significant psychological imbalance resulting in traumatic stress.

12.2.1.2.2   Isolation

From the social and family point of view, the Zika epidemic in the affected communities was disastrous. This led to a deterioration of the social fabric, such as the dissolution of parent-child associations with microcephaly, the breakdown of certain family ties, and the isolation of the mother, children, and their closest relatives. In fact, one of the consequences of this situation is the isolation experienced by mothers and their children, who had minimal assistance from government bodies.

Volunteer teams noted that the feeling and loneliness generated by a couple's losses sometimes leads to risky behavior. In some cases, the self-medication with alcohol, or the use of sleeping medication, was increased. Isolation may also be related to certain difficulties faced by people who have withdrawn in themselves instead of trying to maintain their social networks. This self-exclusion limits the chances of people suffering from depression or anxiety for their new status as a mother with a child with microcephaly, to express themselves and to better manage these difficulties. One mother told us that:

> "We really do not want to socialize with others in the Mothers Club, we all share the same story and that's what we get when we meet. I personally find that this often increases my moral suffering."

The sense of self-exclusion leads to a growing lack of information about the virus as well as next steps to address the end results of the epidemic. The impact on the child, what the future holds for the well-being of the child and the family, and the immediate next step after the child's birth.

12.2.1.2.3 Disintegration of Social Cohesion: Tensions and Guilt

According to some expressions in the focus groups, there were many associations in the community before the Zika epidemic, and social ties were quite strong. Existing community solidarity groups have disintegrated and mutual relationships, which have brought relief to people in difficulties, have deteriorated significantly, increasing the isolation of people in the home and aggravating their distress.

> "When we learned that he was going to be microcephalic we stopped attending church and visiting family, this problem (the baby being microcephalic) was caused by sin."

The way the community handled cleaning and fumigation in the creeks and open water sources, as well as giving out information gave way to a sense of guilt in the neighborhoods. The people who were interviewed stated that they were against the intervention of Red Cross volunteers in their neighborhood at the beginning of the crisis, and sometimes felt guilty that they caused the mosquito breeding grounds. Therefore, they now think they are responsible for the loss of children with microcephaly. A lady told me:

> "We have been reluctant and we have been opposed, for a long time, to sanitation teams involved in our community doing clean-ups and things. We are not dirty people, we are poor."

Disagreements among neighbors about how to deal with cleanliness to prevent the epidemic, especially if they comply with the recommendations of Red Cross volunteers, contribute, with hindsight and analysis of the situation, to create tensions between the neighbors. Some blame others for their attitude during the peak period of the Zika epidemic that may have contributed to the deterioration of the situation.

12.2.1.2.4 Impact of Tradition and Religion

Healers and shamans (espiritistas) recognize that they have some responsibility for poor management of the situation and the extent of the epidemic. This analysis, a posteriori, reflects that this was probably due to his ignorance of the disease.

A week ago in a mothers' *session*, an expectant mother asked:

> "if my son goes out with microcephaly, what medicine do I have to give him to get cured?"

During the exchanges with pastors, religious leaders, and healer it indicated the need to be better informed and trained to include recommendations in their consultations the practice of psychosocial interventions and cleaning and eradication activities of hatcheries to avoid similar spreading of the epidemic. A comment after a training session, echoed by most participants was:

> "We teach and preach in our church that any social or spiritual ill is the work of the devil. We have been telling the parishioners that the Zika epidemic is due to sinful behaviors of the community. We didn't realize today after the Red Cross training that the epidemic was caused by a

mosquito, and water accumulations in places in the community. We can
handle this. It is not so much sin as it is not taking care of our bodies, our
houses, and our communities."

Their speech was sometimes more radical. They speak of the impact of
the "beyond" on the microcephaly situation in all human beings. That is,
what sin have you committed here on earth, that will be punished in heaven.

All these attitudes, the divergences pointed out, contributed to create
division in the community. This assessment has further deteriorated community confidence with regard to healers who made *recommendations* that
affected the health of those affected (mothers and seniors) by the Zika virus.

The divisions in the communities intensified, during the epidemic
crisis, based on religious affiliations. Prior to the Zika epidemic, the Christian community was a minority, they were not widely accepted by the rest
of the community due to differences and contradictions through religious
practices and customs with other members of the communities. During the
greater period of the epidemic, these tensions reached their height, especially within the apostolic movements.

As we let to understand it, Christian groups applied the hygiene
measures brought by Red Cross volunteers and raised the awareness
of members about the need to implement them. Therefore, healers and
shamans accused Christians of collaborating with humanitarian workers
to spread the virus in their village. The president of the Christian minority
in the village told us:

"What happened in this village was very hard. People thought there
were witchcraft problems watching some mostly young mothers get sick
and show symptoms of Zika."

With the changes in seasons from rainy to dry, mosquitoes were not as
prevalent and the Zika cases were reduced. The crisis between groups of
neighbors of different religious orientation also reduced. There was more
psychoeducation and less mosquitoes. They realized that taking care of the
vectors was more a neighborhood social issue that affected everyone and
not a religious issue.

The tensions created by these attitudes, seem to have contributed to
reinforcing a sense of guilt among the affected population, prominently the
mothers, lamenting that they did not follow recommendations that could
have limited the growth of the mosquito, and, therefore reduce the Zika

epidemic. In communities where cleanup activities were planned with the affected community, greater collaboration was achieved between religious groups and community healers.

12.2.1.2.5   Stigma Before the Affected Population

The scale of the Zika epidemic and the lack of accurate information about its origin or results in children within the affected communities generated a sense of stigma. The four mothers with children with microcephaly, which we interacted, expressed feelings of rejection, either from their partner, some relatives or from the community. The four mothers reported that at some point they felt that the comments of family members and community members caused a sense of self-blame, increased anxiety, depression, and the desire to harm themselves and their children. What helped them in those moments was the support of their relatives, volunteers, and health-care staff.

12.2.1.3   DESCRIPTION OF THE MINIMUM NORMS OF APPROACH IN THE MENTAL HEALTH AND PSYCHOSOCIAL SUPPORT OF THE INDIVIDUALS, PEOPLE, AND COMMUNITIES ACCORDING TO THE INTERNATIONAL GUIDELINES

The Minimum Standards for Humanitarian Response of the Sphere project are based on the Humanitarian Charter and the Code of Conduct and are observed by the Red and Red Crescent and NGOs. The Humanitarian Charter is the ethical and legal basis for the four principles of protection, the essential norms and minimum standards established in the Sphere project. The Charter places special emphasis on the principle of humanity and calls for "all state and non-state actors to respect the impartial, independent, and nonpartisan role of humanitarian agencies..." (Humanitarian Charter, § 3). The term "nonpartisan" is used deliberately (instead of the term "neutral") to allow for different interpretations of the principle of neutrality. Echoing their use in the Code of Conduct for the Red Cross/Red Crescent and NGOs, the intention behind the expression "no partisans" is to reflect the concept that humanitarian actors should not take sides between warring parties. The principle of impartiality is also dealt with extensively in Sphere Protection Principle 2.

Other series of standards in mental health and psychosocial support are proposed by the Federation (IFRC), which since 1991 recognizes the psychosocial support program (PSP) as a transversal program under the Division of Care and Sanitation. The Federation Reference Framework 2005–2007 defines psychosocial support as helping people to recover from a critical life-altering situation, in this case, the epidemic caused by the Zika virus. Recently, the WHO published a guide entitled "Psychosocial support for pregnant women and families affected by microcephaly and other neurological complications in the context of the Zika virus." The latter document provides general guidelines that are operationalized at the local level.

Epidemics, such as those caused by *A. aegypti*, impact the psychosocial status of those affected, infected, and social networks around them. It is, therefore, important to address psychosocial needs within the framework of humanitarian response. Psychosocial support is a transversal intervention of the guidelines of the SPHERE project and that the guidelines also use them. The essential norms relate to people's social and emotional well-being and their ability to help themselves and recognize that it is important to design the response in a way that respects the right to live with all dignity.

The evaluation standard pays more attention to the assessment of the immediate and more general context, the capacity of affected communities in relation to the response and the psychosocial effects of the epidemic. Sphere emphasizes the importance of self-help in the community, concrete support for networks, community projects and the need to include individual and community initiatives in the response and recovery process.

The project Sphere (2011) introduces four important issues for psychosocial support: (1) protection principles suggests the protection of individuals and communities from physical and psychological damage caused by violence and cohesion (Sphere 2011, p. 38–43); (2) fosters the need to *harness* the capacities of communities to increase self-help (Sphere 2011, p. 43); (3) all psychosocial aspects as a formal crosscutting theme in all guidelines; and (4) the health standard includes psychosocial support such as self-help and community social support, psychological first aid, the importance of primary health care, and the formulation of plans propels a community health system (Sphere 2011, pp. 335–336).

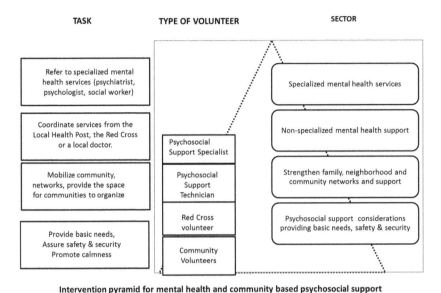

Intervention pyramid for mental health and community based psychosocial support

FIGURE 12.1   Intervention pyramid, personnel to serve the tasks of the response, and the titles of the interventions at each level.

Epidemics caused by vectors, such as those caused by the *A. aegypti* mosquito require psychosocial support at different levels. This complementary support system covers the needs of diverse groups including pregnant women, women and men of childbearing age, children with microcephaly, family members, neighbors, and the community.

The *first level* of the pyramid of intervention occurs in the affected community. The first level is related to basic health satisfaction and protection through the control of epidemics and the elimination of epidemics that cause communicable diseases such as Zika. The volunteers are responsible for community psychoeducation, reduction of rumors and stigma, and promulgating calm in the community. There are several examples of this activity in all sites visited and the diverse approaches used.

In the first step, activities are promoted that increase psychosocial well-being and strengthen security in the family and the neighborhood of mutual support. In one school, a group of children received guidance on the school site and then participated in a community campaign to eliminate mosquito breeding sites.

In integrating psychosocial support in relation to the Zika epidemic, the volunteers consider the activation of community networks, to provide psychological first aid, begin to offer psychoeducation in schools and community, and if necessary refer some people to the community health unit or post, or to the regional hospital to receive no specialized support.

Community volunteers should be able to offer basic psychological first aid and emotional support to those infected and affected. Community volunteers have a basic knowledge of how to carry out community screening, determine community resources, and draw a rustic map of housing, and an annotation of where the infected and/or affected persons are. In one of the countries they described the psychosocial support intervention was as follows:

> *"community volunteers visit the community and identify people who feel sick (fever, rash, pain all over the body). If people show symptoms of the Zika virus, they are referred to the community health post. The health post will assess the person, and if needed sends the person to the regional hospital for specialized services. Once the person returns to the community, the volunteer activates a network with the neighbors and ensures that the sick person has a support network of care. Usually the Red Cross must* identify one *community volunteer in every street or sector"*

Community facilitators may be midwives, public health technicians, or rescuers. In one of the sites, volunteer first-aiders from the Red Cross of the Red Cross youth canvas the communities and refer neighbors who feel sick to the Community Health Unit. There are already protocols of services between both groups and sometimes *both* groups leave to visit the immediate community. They come together to carry out community-based preventive health activities. On the visit, they were planning on cleaning of a stream that crosses the community to eliminate mosquito breeding sites.

Red Cross community facilitators have the basic skills of community volunteers; they also learn to develop a profile of community psychosocial needs, advanced psychological first aid, information and communication techniques, and community development. They carry a toolkit with activities for children and psycho information for marginalized populations. Each Red Cross community facilitators support approximately ten community volunteers.

In another site, Red Cross volunteers carry out vigilance in target communities and refer people with fever to the health post for further evaluation. On returning home, volunteers mobilize neighbors to support the person that were sick. In this step, the skills of psychological first aid are used to make the population feel secure, connected, calm, and optimistic. The activities are structured to offer social, physical, and emotional support. These activities promoted the sense of usefulness by helping oneself and the community.

One strategy that was very well used in all sites was engaging "spontaneous volunteers" in "red areas" to carry the message about the Zika epidemic to these populations.

The *second level* of the pyramid of support includes community psychosocial support technicians. This level suggests specific support to recover the protective factors provided by family and community support systems. This level reestablishes the lives of those affected or infected through planned activities in social networks, cultural or artistic centers, churches, various groups, formal or informal education, or support groups for fathers and mothers of children with microcephaly. Support activities are conducted to help parents, grandparents, and other caregivers cope with their own stress.

An example of this activity was experienced in one of the community visits, where the extended family cares for and supports a child with microcephaly and his mother, the community school has developed a support plan for the child. Public health agencies and the rehabilitation hospital collaborate to provide early stimulation to the child. Finally, the community is provided with accurate information that has defined a strategy of constructive coping.

A female Red Cross volunteer was strategically inserted in the group of community leaders, mostly males; this facilitated the development of a Mothers' Club. On the day of our visit, she led the group and group *dynamics* that promoted physical relaxation. The Dominican Red Cross has developed a methodological guide to prepare technicians in crisis intervention that contains chapters in psychosocial support and psychological first aid that can easily be used to train community psychosocial support volunteers and technicians. These volunteers have some preparation in psychological first aid and peer-to-peer counseling.

The *third level* of the pyramid is made up of specialists in psychosocial support. The third level represents support from social workers, volunteers, and other staff at the health post for mothers and fathers affected by the Zika virus and have children with microcephaly. These interventions include psychological first aid and psychosocial groups structured for mothers or fathers. In one site the Mothers' Clubs at the Red Cross Hospital learned how to do relaxing activities, formed a network of support using their cellular phones, and received psychoeducation.

Red Cross specialists were recruited by the Red Cross. This is a professional workgroup consisting of psychologists, social workers, counselors, or cultural anthropologists. They must also have a minimum of three years of experience as Red Cross volunteers. They are part of a team of four people serving in a time of crisis or disasters. They have experience in training volunteers in the first three levels, preparing materials, taking care of the needs of volunteers, serving as community organizing experts, and ensuring that all volunteers psychosocial support are adequately met.

They are trained in psychosocial support guidelines (Sphere and MHPSS) and can adapt to different types of disasters. According to accepted practices, in the time of disaster, the national psychosocial support team becomes an adjunct to the Department of Emergency Management and the Ministry of Health responsible for managing the psychosocial component of a disaster and assures that no harm is done to the survivors.

The *fourth level* of the pyramid is made up of specialized mental health services. The fourth level refers to a small group of people who usually have difficulty coping with the suffering caused by pregnancy and at birth with a child with microcephaly. Sometimes the reaction to the sequelae of the Zika virus is caused by preexisting mental health conditions. Therefore, referral to mental health experts is required, who can diagnose a disease and use clinical methods to make the person feel better. These experts are psychologists, psychiatrists, psychiatric nurses, social workers and/or any other person either in Government Hospitals or in private practice who can support a person with mental health needs, and present symptoms that can be diagnosed using The Diagnostic and Statistical Manual of Mental Disorders, Fifth Edition (DSM-5) or The International Classification of Diseases, Tenth Edition (ICD-10).

Psychosocial guidelines have been developed by the Psychosocial Support Reference Center (Providing Psychosocial Support During Epidemics) aligned with the Inter-Agency Standing Committee (IASC)/ Mental Health and Psychosocial Support (MHPSS) international guidelines. These guidelines are intended for people working in mental health and psychosocial support and provide several recommendations for psychosocial actions that correspond to the basic areas of humanitarian work, such as protection, health, and education. These basic themes are included in a response-oriented response to the sequelae of the Zika virus.

12.2.1.4   DESCRIPTION OF INTERVENTION FOR PSYCHOSOCIAL SUPPORT OF THOSE AFFECTED/INFECTED

Psychosocial support interventions are part of a critical component in Zika epidemic preparedness and response activities. This section will discuss two components: (1) community mobilization and (2) psychological first aid.

12.2.1.4.1   Community Mobilization

Community mobilization is a tool that permits community volunteers and Red Cross staff to establish and strengthen links with affected communities in ways that increase the psychosocial well-being of the community. The most frequently used psychosocial activities (that can be measured in a logical framework) are:

- Hygiene and health promotion campaigns that focus on sensitizing the population to the transmission and treatment of the Zika epidemic.
- Use of social media to provide accurate and immediate information on the Zika epidemic.
- Use of the media, radio, television, community theater, posters, and drawings to carry messages that reduce stigma and allow the inhabitants of communities to express their doubts or concerns.
- Focus groups that promote learning among neighbors, allow free dialogue between members of the community, and alleviate fear, loss, and feeling ashamed.
- Meetings with faith groups, healers, and community leaders to disseminate sound messages about the impact of the Zika epidemic and to provide adequate support to people affected and infected by the virus and mothers of children with microcephaly.

12.2.1.4.2   Psychological First Aid

Psychological first aid is immediate, experiential, and simple activity that encourages calm and assure affected populations that their reactions are normal, in the face of an abnormal situation. Some of the common techniques of this intervention are:

- Provide clear and consistent information. In providing this information we have to be clear that people can demonstrate their fears, their frustration, and anguish. These behaviors are not necessarily an expression of opposition to the person or the message, but rather shared emotions. Be aware of the rumors that are floating around the community, so that clear messages can be dispelled.
- Link with appropriate services. Affected and infected individuals may have specific needs to meet their basic needs, or what are the steps to access services needed to resolve their condition. Keep vigilance of people who are exhibiting severe stress reactions, are provided with protection and are referred to specialized services.
- Connect affected or infected people with loved ones and community networks. Psychological first aid helps affected people identify those individuals, institutions, or networks that can help them reintegrate into their communities.

12.2.1.5   HOW THE FINDINGS FROM THIS STUDY ASSIST THE RED CROSS IN INTEGRATING PSYCHOSOCIAL SUPPORT INTO A RESPONSE TO AN EPIDEMIC

The integration study had two objectives. The first was to review the existing documents in psychosocial support and to determine how they apply to the Zika prevention and response. Existing materials were examined to determine if the information was clear and consistent. From the beginning, the study the researcher was aware that unlike other sectors, psychosocial support is not composed of a series of formulas and protocols. Working with manifest emotions and attitudes at the community level requires that the volunteer learn basic techniques of community mobilization that were culturally, linguistically, and contextually appropriate. Using materials translated from other languages and other contexts did not prepare volunteers nor helped communities affected by the Zika epidemic.

The second objective was to visit sites in the four countries (Brazil, Haiti, Puerto Rico, and Santo Domingo) and document the specific activities of psychosocial support using the guidelines from the IASC/MHPSS as the benchmark. We wanted to engage in conversations with end users about their needs and what they thought were best ways to alleviate suffering in their communities. The preferred mode for data collection was focused groups. The challenge with this approach were the languages involved (Creole, Spanish, and Portuguese).

The integration study was interested in exploring how the Red Cross contextualized international guidelines so that they meet local needs. For example, we note that all sites used the pyramid of services advanced by the IASC/ MHPSS guidelines. It is interesting to note that volunteers referred to level 1 and level 2 intervention. Volunteers with degrees (i.e., psychologists) viewed the role of the Red Cross at level 3 and level 4 of the Pyramid; but all started with the fourth level. One possible explanation is that the clinical model was used to recruit the people who were to carry out the psychosocial support project, and therefore they should be psychologists. In a single National Society visited, we noticed that the principles of community mobilization were used.

12.2.1.6   EXPERIENCE

12.2.1.6.1   Field Experience

Each of the National Societies visited is culturally and contextually unique. Its modus operandi has a local flavor but the principles and components are similar. Psychosocial support is a crosscutting theme for disaster response in line with the response guidelines presented by the IFRC. During the visit, we inserted a "problem" that required a working knowledge of the relationship between the staff of the Red Cross and their knowledge of all the parts necessary to solve a "psychosocial" problem. It was rather interesting that psychosocial support had different meanings for different people, also those in charge of psychosocial support in the National Society and the Branches were volunteers, while the Zika project personnel were paid staff.

12.2.1.7   TOOLS FOR DATA COLLECTION

During the visits, the same measurement qualitative techniques were used in all the sites. The qualitative evaluation assisted the researcher to determine the level of distress of pregnant mothers and mothers with microcephaly, family members, affected community members and medical staff in health posts, and/or hospitals and clinics of the Red Cross. The purpose of these qualitative assessments was to determine feelings, actions, stressors, and strategies for advancing and reaching the psychosocial competence expressed by affected and infected and Red Cross volunteers.

12.2.1.7.1   Focus Groups

This activity explored the feelings, beliefs, and attitudes of the groups of mothers, families, and members of affected communities. A moderator facilitated the structured group process through which mothers, family members, and community members discussed the psychosocial support activities used and future needs of children with microcephaly. Participants in this activity provided insights, experiences, beliefs, knowledge, and attitudes about how the community can achieve psychosocial competence after the Zika epidemic. The result of this activity provided information about the strategic planning process, validated other findings and in the future will help in activities focused on collaborative solutions. For example, a focus group in one of the sites discussed the importance of a Community Medical in a remote neighborhood.

12.2.1.7.2   Participant Observation

This method allowed data collection by selectively observing and monitoring the behavior of members of Mothers' Clubs and volunteers and their interactions within their group and with those outside their group. Observation *served* as a tool to support other sources (i.e., focus groups or interviews) by confirming or complementing known data.

12.2.2   NEEDS IN MENTAL HEALTH AND PSYCHOSOCIAL SUPPORT OF INDIVIDUALS, FAMILIES, AND COMMUNITIES IN RELATION TO ZIKA

It is impossible to measure mental health and psychosocial support needs as National Societies do not have the staff or the psychometric resources to perform these measurements. However, based on the qualitative evaluation as a part of the information collected for this study, we can surmise that the general population needs accurate, clear, and time-sensitive *information*. Pregnant women in areas where Zika cases were identified had fear, anxiety, guilt, and sadness for what can happen to them. We noticed in some meetings that mothers and some volunteers showed psychosomatic reactions such as physical pain, lack of appetite, hyper vigilance and some reported being unable to sleep or experience nightmares related to the Zika work. Volunteers and some members of the community were affected by the constant information through the media or rumors expressed during visits to the community. One of the participants in the mother's group asked what type of pills she would have to give to her child with microcephalia.

12.2.3   IDENTIFIED RESOURCES USED TO IDENTIFY NEEDS

National Societies used volunteers to disseminate clear and accurate information, maps were developed in communities that targeted pregnant mothers and/or persons reporting having contracted the Zika virus. Volunteers, through visits, provided referrals from people to health posts, conducted orientation meetings and participated in psychosocial support *meetings* with mothers of children with microcephaly. Among other identified resources, National Societies mentioned psychoeducation activities to help volunteers and community members understand that reactions to the Zika epidemic were normal reactions to abnormal situations.

Globally, there are six activities that volunteers indicated them in relation to the response to the Zika epidemic.

- Increase the capacity of volunteers and staff in psychosocial support. Psychosocial support is a cross-cutting activity in the repertoire of interventions supported by the Red Cross. All volunteers will benefit from capacity building in psychological first aid, early understanding of psychosocial reactions, and self-care.
- Enhance the capacity of Red Cross staff to carry out activities at the community level. The volunteer will benefit from capacity building in the process of planning and implementing community psychosocial responses. These include understanding (1) what psychosocial

well-being is, (2) how it is identified, (3) what tools community
needs, (4) who provides psychosocial support activities, and (5)
how we know the community has solved it their psychosocial needs.

- Community-level support by mental health experts (psychologists
  or psychiatrists). The responsibility of experts in mental health
  is found in private and public hospitals and clinics. Where these
  people can identify clinical sequelae and appropriate treatments
  (therapy, counseling, and/or medication). The field staff responsible for implementing the psychosocial response in the communities must be volunteers, technicians, or specialists in community
  psychosocial support. This suggestion is consistent with the practices pre-established by the Ministries of Health in the countries
  visited.

- Neighbors in communities are considered key actors in psychosocial
  activities. The activities in the National Societies visited are limited
  to offer the resources to eliminate the mosquito of *Aedes aegypti*.
  Meetings with Mothers' Cub take place over a period of several
  weeks. Handout and other visual didactic materials are shared
  during these meetings. We realized the importance of community
  empowerment of psychosocial activities. This event empowered
  and acknowledged the community network for its psychosocial
  well-being. By empowering neighbors to be key players and active
  in psychosocial activities, the space for a more resilient community
  opens.

Two factors in the planning and implementation of psychosocial
support in communities affected and infected with the Zika virus were:
(1) communication and (2) nature of the information. Communication
considers cultural sensitivity and information is accurate and contextual.
The volunteer must learn and understand the use of popular media in the
target community. You should plan written, visual, or spoken messages
accurately and effectively always.

The focus of community psychosocial support has to be community and school. Community mobilization focusing on the psychosocial
support is inclusive and global. Everyone participates in the planning of
interventions through mapping exercises. Psychosocial support activities
in schools are developed through the existing classroom support system,
health care, child care, and social welfare of all and all students.

12.2.4 TRAINING MECHANISMS TO STRENGTHEN THE CAPACITIES OF VOLUNTEERS

Training mechanisms are developed as per the National Society response methods to disasters and the place they attribute to psychosocial support before the organizational schemes of the National Society. Each of the three National Societies has a different approach to staff capacity building.

In another country a National Technical Committee has been organized on the basis of IASC materials, the Sphere project and the guidelines of the Reference Center in Copenhagen, other local materials, as well as the experiences of the volunteers to form a training strategy to develop the capacity of volunteering.Guatemala, on the other hand, has training mechanisms and a strategy to develop the capacity of volunteers at different levels of responsibility for psychosocial support activities. In another site, the National Society recruited a psychologist as a psychosocial support technician. This person has targeted some volunteers and prepared them on psychosocial support approaches and techniques. At the volunteer meeting, which I attended, there were a plethora of characters: volunteer first aiders, high school students doing their practice, members of the Red Cross Youth.

In the four countries visited, volunteers identified a problem of the fear of entering "red zones," although they recognize that these are the places with the greatest need. One of the focus groups identified their needs related to training in psychosocial support, while another group was capable to identify "community mapping" steps "roughly." However, interviewees had difficulty in identifying psychosocial problems and using Red Cross resources to refer to these cases.

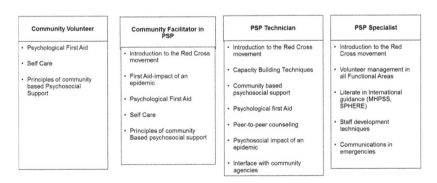

Continuum of Capacity Building in psychosocial support

FIGURE 12.2   Continuous training of psychosocial support personnel.

As participants in the focus groups presented their inputs, we noticed that, depending on the volunteer's task, the training needs were presented. Figure 12.2 presents a synopsis of the continuum of capacity building in psychosocial support as identified by volunteers in one National Society.

The country's contingency plans include the Red Cross as responsible for providing psychosocial support in an emergency. In addition to responding to the Zika epidemic, they also provide services to schools, the Child-Mother Hospital, the shelter for young girls, and are currently providing support to areas where there is armed conflict. Volunteer staff and the National Society see the role of psychosocial support at levels 3 and 4 of the service pyramid (WHO, IASC, and IFRC). Therefore, the training of volunteers has been intensive and extensive.

In summary, during the visits, we could identify strengths and weaknesses. Strengths include the use of mental health and psychosocial support as part of emergency and disaster responses and a large number of psychologists who use branch buildings and clinics to serve mostly privately-owned children. Among the difficulties, there are capacity-building gaps in the first two levels of response that include community psychosocial support. This can be eliminated if the Zika–PSP interface pursues three objectives: (1) to develop the capacity of national societies by implementing epidemic eradication programs that increase community well-being; (2) increase the capacity of volunteers (rescuers, community members, and youth) to respond to the urgency of social and psychological needs caused by the epidemic; and (3) increasing support for education of volunteers on the psychosocial impact of an epidemic such as Zika epidemic.

12.2.4.1 NATIONAL SOCIETY ORIENTED

National Societies addressed the task of providing psychosocial support to affected communities and within them to individuals who experienced the Zika virus, pregnant mothers, mothers with children exhibiting microcephaly characteristics, and families and schools. In the three countries visited, attention was given to marginalized populations and strategies were developed to provide services in the red areas. There was a deliberate effort to build the capacity of members of the marginalized communities and engage them as spontaneous volunteers.

Rapid response strategies were used at times. It takes some time to understand that the Zika epidemic is a vector-borne, low-impact, but long-lasting disaster. It was, therefore, necessary to readjust the psychosocial support response to (1) focus on community-level stress reactions, that included anxiety, sadness, guilt, stigma, and fear, (2) focus on psychoeducation, and simple, accurate, and contextualized information, and (3) provide capacity building to volunteers in subjects such as self-help, psychological first aid, and community psychosocial intervention strategies. These activities should be further enhanced in the response strategy of the National Society.

One of the elements identified that require strengthening during the response, is the empowerment of affected and infected communities. National Societies should also continue to strengthen their networks with government agencies (Ministry of Health), and with nongovernmental agencies, community and faith-based organizations and other groups that are willing to collaborate in the psychosocial development of communities.

An element that is missing and should be incorporated into the response, should be the emphasis on the preparation of National Society staff (self-employed and volunteers) in self-care techniques, peer-to-peer counseling, psychological first aid, and specific emotional support techniques.

In terms of capacity building of response personnel, it should be timely and practical. All training should be practical, supervised in the field, and be part of a training ladder. The materials of different centers exist in the media. These should be translated using a committee of experts in the field, contextualized to the reality of each country, and written precisely, clearly, and easily understandable by people who have not completed higher education.

12.2.5 RECOMMENDATIONS

We recommend that (1) National Societies develop a practical policy that includes psychosocial support as an element of approach in responses to disasters, emergencies, and low-intensity events such as the Zika epidemic. (2) The Secretariat take into account the different phases of support to National Societies and inter-institutional coordination related to community psychosocial support, considering the national linguistic, cultural, and contextual to the phases one and two of the pyramid management of the program, (3) implementation activities including psychosocial support activities at all levels, (4) training of volunteers as a progression of knowledge, (5) monitoring, and (6) evaluation.

We recommend that the efforts that have been developed as part of the response to the Zika epidemic continue to be promulgated by the region and through related organizations to develop a community psychosocial support strategy that fits epidemics and the consequences of long duration that cause them.

12.3   CONCLUSION

This chapter shares with the reader the lessons learned after working in the field of mental health in disasters and in psychosocial support for more than 37 years. It is difficult to translate into full details of both programs, their similarities and how each approach is used to reduce suffering in the affected population. The Red Cross movement has specialized mental health teams that address physical and emotional needs in major disasters. Psychosocial support teams at National Societies combine lifeguards with health personnel to form their teams that support people in emergencies such as the epidemic caused by the Zika epidemic.

The material contained in this monograph should be sufficient for educators and reference centers to develop strategies and curricula for local psychosocial teams that will provide services to people affected and/or affected by the Zika virus. We trust that the movement has enough tools available to prepare psychosocial support teams. What is needed, and this writing cannot provide, is instructors and writers of curriculum that transfer the qualitative knowledge so that training is transferred to practice.

REFERENCES

American Red Cross. *Afrontar problemas en el mundo hoy;* American Red Cross: Washington, D.C., 2011.

CREPD. *Curso de apoyo psicosocial en emergencias;* IFRC/CREPD: San Salvador, El Salvador, 2015.

CREPD. *Módulo de formación del virus* Zika*: Manual de Referencia;* IFRC/CREPD: San Salvador, El Salvador, 2016.

Cruz Roja Americana. *Afrontar problemas en el mundo de hoy: Primeros Auxilios Psicologicos y capacidad de recuperación para familiares, amigos y vecinos;* American Red Cross: Washington, D.C., 2010.

European Federation of Psychologists Association. *Lessons learned in psychosocial care after disasters;* Council of Europe: Strasburg, Germany, 2010.

IASC. *IASC Guidelines on Mental Health and Psychosocial Support in Emergency Settings: Checklist for Field Use;* IASC: Geneva, Switzerland, 2010.

IFRC. *Ayudando a sanar: Manual de voluntarios;* IFRC: Clayton, Panamá, 2008.

IFRC. *El cuidado de los voluntarios: Conjunto de herramientas para el apoyo psicosocial;*IFRC Psychosocial Support Center: Copenhagen, Dinamarca, n.d.

IFRC. *Lay Counseling: A Trainer's Manual;* IFRC Psychosocial Support Center: Copenhagen, Denmark, n.d.

IFRC. *Intervenciones psicosociales;* Centro de Referencia para el apoyo psicosocial: Copenhagen, Dinamarca, 2009.

IFRC. *Caring for Volunteers: A Psychosocial Support Toolkit;* Copenhagen, Denmark, 2012.

IFRC. *Lay Counseling: A Trainer's Manual;* IFRC Reference Centre on Psychosocial Support: Copenhagen, Denmark, 2012.

IFRC. *Health Emergency Response Units-Psychosocial Support Component-Delegate Manual;* IFRC Reference Centre for Psychosocial support Centre: Copenhagen, Denmark, 2012.

IFRC. *Providing Psychosocial Support During Epidemics;* IFRC Reference Centre for Psychological support: Copenhagen, Denmark, 2016.

Organización Mundial de la Salud (OMS/WHO). *Apoyo psicosocial para las embarazadas y las familias afectadas por la microcefalia y otras complicaciones neurológicas en el contexto de virus* Zika; Organización Mundial de la Salud: Ginebra, Suiza, 2016.

Pan American Health Organization. *Mental Health and Psychosocial Support in Disaster Situation in the Caribbean.* Washington, DC: PAHO/WHO, 2010.

Pan American Health Organization. *Zika. InformeEpidemiológico de Guatemala.* Noviembre 3, 2016. PAHO/WHO: Washington, DC, 2016a.

Pan American Health Organization. *Zika Epidemiological Report Brazil.* PAHO/WHO: Washington, D.C., 2016b.

Prewitt Díaz, J. O. *Primeros auxilios psicológicos*; American Red Cross: San Salvador, El Salvador, 2002.

Prewitt Díaz, J. O.; Morales, B. *GuiaMetodologica: Diplomado para interventores en crisis*; Cruz Roja Americana: Guatemala, 2003.

Prewitt Diaz, J. O.; Dayal de Prewitt, J. O. Sense of Place: A Model for Community Based Psychosocail Support Program. *The Australasian Journal of disasters and trauma studies.* **2008**,*2008*, 1.

Prewitt Diaz, J. O.;Bordoloi, S.; Miashra, S.; Rajesh, A. *Community Based Psychosocial Support Training for Community Facilitators;* American Red Cross and Indian Red Cross Society: New Delhi, India, 2007.

SPHERE. *Humanitarian Charter and Minimum Standards in Humanitarian Response;* The Sphere Project: Geneva, Switzerland, 2011.

UNICEF. *Inter-Agency Guide to the Evaluation of Psychosocial Programming in Humanitarian Crises*; United Nations Fund: New York, NY, 2011.

CHAPTER 13

PSYCHOSOCIAL SUPPORT IN THE COMMUNITIES: A REVIEW OF EVIDENCE INFORMED PRACTICES

JOSEPH O. PREWITT DIAZ

Chief Executive Officer, Center for Psychosocial Support Solutions, Alexandria, VA 22310, USA

CONTENTS

13.1 INTRODUCTION

A catastrophic event brings about individual emotional reactions for the affected people as well as external reactions that impact the neighborhood and community. The Sphere Project (2011) suggests that in order to better address the emotional needs of affected people, mental health and psychosocial support interventions are most appropriate. These two approaches complement each other. The mental health approach includes institutionalized attention (clinic, private office, or hospital), medication, and individual therapeutic approaches that alleviate the stress caused by the disaster or humanitarian crisis and promote calmness.

The other approach is psychosocial support, that is, a consideration of how people are impacted by environmental changes taking place in the community, their perceptions based on cultural interpretations, and how social and economic conditions impact the behavior of the affected people. This intervention includes engaging the affected people in a dialogue that identifies their needs. Psychosocial support uses activities that promote the social and psychological well-being of the affected population that considers the mutuality of all of the affected people in the identification of needs, possible interventions, and agreed-upon results. These activities include community mapping or initial assessment, defining culturally and contextual interventions, identifying the existing community-wide social capital that may join the community as resources, and establishing ways to monitor the progress of the community.

Most of the chapters in this book use descriptive methods or case studies to explore planning, development, monitoring, and evaluation of diverse types of psychosocial programs in diverse settings. This chapter will (1) provide a review of important methodological issues involved in cultural and contextualized programs that identify threats to the validity of the projects in a variety of cultural, linguistic, and contextual settings, (2) briefly revisit the examples provided by several authors in this book as well as examples from other literature, and (3) propose best practices for implementing psychosocial support programs (PSPs) in multiple settings.

This chapter draws upon the expertise of the authors of this book and those found in the literature to develop a three-stage framework for a program: (1) develop community-based psychosocial support based on the principle of mutuality, (2) the alignment of interventions with culture and context, and (3) validation of materials and educational tools that are culturally and contextually appropriate. The goal is to identify best-practices recommendations for clinicians and humanitarian practitioners.

13.2   DEVELOPMENT OF INTERVENTIONS

Community members have been engaged in assessing the psychosocial support needs. Among the target population, we would find those with pre-existing mental health issues, those who have psychosocial disabilities, and all other sectors in the community. The community members identify what constitutes a psychosocial program for them, they define their cultural context, and how the principle of mutuality serves as a tool to include all community members (within), and external stakeholders (applicability of program components across external groups). This stage further explores how to operationalize the psychosocial interventions.

Psychosocial support consists of any type of program of local or outside support that aims to protect or promote psychosocial well-being (IASC Reference Group, 2017). We want to learn what the people from within a culture think about their understanding of psychosocial support, and how they can use these activities to handle individual psychological concerns and group, family, neighborhood, or community interactions that may be causing stressful situations.

During the early disaster mental health (DMH) experiences overseas, the target population and the donor population were not quite convinced of funding programs where they could see and count tangible objects. This presented significant challenges to the first generation of DMH delegates for the American Red Cross. Two examples are appropriate to represent our point. The first example occurred in India. The American Red Cross proposed a program after the Bhuj (Gujarat) earthquake to the Indian Red Cross. This was a signature program within the American Red Cross that focused on providing psychological first aid, short individual intervention, psychoeducation, and brief psychosocial support to people affected by the disasters. The program would address all the affected areas in the state of Gujarat and at the same time leave a cadre of well-prepared professional, para-professional, and community volunteers capable of attending mental health needs through the recovery and longer terms needed.

At approximately the same time communal riots in Ahmedabad created great turmoil among the Hindu and Muslim population. A group of Indian mental health professionals understood the need of the people as a collective need for harmony and supported what was defined as psychosocial care at the time. The focus of both approaches was to alleviate fear and promote calmness. The way of delivering (tools used) was different. The American Red Cross diverted the funding to support Project Harmony in Ahmedabad, a psychosocial care program that lasted for three years and provided assistance to thousands of people, mostly from the Muslim population. The program did not serve the needs of the total affected population, which was comprised of Hindu, Muslim, Christians, and others.

The second example occurred in Central America, where after the El Salvador earthquake of 2001, a proposal was submitted to a US Federal Agency seeking funding for a DMH/psychosocial program in four Central American countries. While the activities and materials were psychosocial support, the funding agency called these activities disaster preparedness and developing resilience in community and schools. It was not until many years after that was developed that the same Federal Agency and the Red Cross entered into an agreement to develop a framework for mental health and psychosocial support.

Transferring technology from one culture to another group or from one country to another is a sensitive process that does not guarantee similar results or success after a given time period. There are three potential ways of transferring a program from one culture to another: using prepackaged programs either from the American Red Cross or from the Psychosocial Support Centre in Copenhagen; establishing consultation with the local national society to determine what activities have taken place and complementing the volunteers with a new set of psychosocial support skills. The third is for the community volunteers to conduct a needs assessment, and with the community participation to develop a culturally and contextualized intervention.

Case Study 1: Dash (2017, in this volume) describes how he contextualized a DMH program by converting the American Red Cross model into the Orissa State Branch model of psychosocial care. He took this model with him to the Republic of Maldives where it took two years to develop a culturally and contextualized program. Among the significant contributions was the development of illustrations and videotape materials that were used intensively in the training program. This mode of development was appropriate for the site because the population of the Republic of Maldives includes many atolls that were not easy to reach. He later took this model to Bangladesh. It took him a year to make the necessary culturally and linguistic adaptation to be able to provide coherent, community organized, managed, and monitored psychosocial support activities.

Dash (2017) in this volume explained how he used the "emic" approach. The emic approach refers to the transference of a concept from one country and culture to another country and culture. The reader must understand this approach for program development looks at international results in a shared cultural or contextual frame of reference, and, therefore the planned activities may not have a conceptual grounding that is readily understood in both cultural and contextual settings.

Case Study 2: The use of the "etic" method in program development considers a broader comparative analysis that involves two or more cultures. The etic mode of program development involves developing an understanding of psychosocial support by comparing predetermined characteristics of the target communities. Dayal de Prewitt (2017, in this volume) explains two examples in detail – (1) bringing psychosocial support interventions for the children and teachers in the Republic of Sri Lanka and (2) utilizing techniques used in two states India. She conducted a series of meetings with a selected group of clients (teachers, parents, and upper elementary school children). As a result, she developed a list of activities that would inform the intervention and the overt behaviors exhibited by children as a result of said interventions. She describes how initially she recruited community volunteers. This group was trained in psychological first aid, community assessment, and simple interventions such as conducting meetings with the neighbors and recording and prioritizing ideas. Through these simple interventions, the volunteers were able to identify proactive behaviors among the population. She was able to develop a PSP that was etic in nature. Two culturally different groups (Sinhala and Tamil) as well as religious groups (Buddhists, Muslims, and Christians) conducted activities unique to their respective groups but yet resulted in an appropriate psychosocial intervention.

This approach was found to be the most practical in terms of time pressures and financial considerations. Berry (1990) suggested that if constructs are perceived to be generalizable across cultures (i.e., root shock, or community, and school interventions), expenditure of resources may be reduced because there will be no need to conduct an emic approach that observes the cultural characteristics in every culture.

Psychosocial support interventions have cultural and context-specific meanings, and these interventions are derived by observation and participation in the target locations. The interventions keep aside cultural bias and become familiar with relevant cultural differences (i.e., language, faith identity, gender, caste, and status within the society). The etic approach addresses cultural groups as opposed to nations. For example, there is a difference between framing psychosocial support as an Indian model or an American model, and the examination of cultural influences within subcultures (i.e., Hindu and Muslim in the Indian context or the Sinhala and Tamil in the Sri Lanka context).

Psychosocial support focuses on the "etic" nature of program development. In these initial stages where the implementers are attempting to understand the cultural groups in a community, the methods used to relate to each group are related to mutual empathy and intersubjectivity (Jordan, 1986). Intersubjectivity carries with it some concept of motivation to understand another's meaning system from their frame of reference and ongoing sustained interest in the inner world of each community group.

A model psychosocial support program that explores the relationship between people (intersubjectivity) community group involved in the psychosocial program development process includes: (1) an interest in and awareness of responsiveness to the subjectivity of one and the other group through empathy (Atwood and Stolorow, 1984); (2) a willingness and ability to reveal the agendas of the funding agency to the groups of affected people and to make the donor needs known to all stakeholders, giving access to both groups (the affected people and the donors) (self-disclosure) (Surrey, 1984); (3) the capacity to acknowledge one's needs without consciously or unconsciously manipulating the other group to gain gratification while overlooking the existing experiences (social capital) (Jordan 1986); (4) valuing the process of knowing, respecting, and enhancing the growth of the other; (5) establishing an interacting pattern that both groups are open to change in the planning process (Jordan, 1991).

Steps in developing a cultural and contextual program are:

- Identify the psychosocial support framework and its components (i.e., community, schools, and specific group in the community).
- Engage in idea sharing among all segments of the community about the objectives and activities of the PSP, consult with members of all community groups, bring the community together in focus groups and work together (emic).

- Generate activities a "sample of convenience" (made up of people who are easy to reach) that responds to every activity suggested by the target community, identifying activities that have the same meaning for the target population.
- Once the etic dimension is identified, develop activities in each group that will achieve the objectives for the PSP.

In program planning process, the cultural sense of place is of great importance and must be considered at all levels and segments of program development. There are least two levels that must be kept in the design: (1) the individual level, where psychological processes, attitudes, and values interface, and (2) the societal level where anthropological, ecological, and economic (social capital) trends often interact.

Best Practice: To consider the appropriateness of the PSP, consideration of culture and context are paramount.

1. Do not transfer a program from one country to another; rather develop psychosocial support among cultural and linguistic groups. Keep in mind the cultural determinants such as language, proximity and topography, religion, economic development, technological development, political boundaries, and climate. For example, language differences were found by Dayal de Prewitt (2017) and Bhatra (2017) in separate cultural groups because they affect how symbols and meaning to life events are communicated (cultural sense of place and solution focused activities). Alessandri and Zoumanigui (2017, in this text) identified how religions are a source of cultural variation within groups of people and within the same country, because of unique traditions and customs. These may bring about behaviors and perceptions of others in the community that become barriers in the recovery process.
2. Incorporate a cultural sense of place as a crosscutting issue in all psychosocial interventions. This consideration may help the implementers to understand the similarities among the target population. Context is important in understanding cultural syndromes (Triandis, 1994) such as complexities or tightness/looseness of given groups within the society. Cultural sense of place is a moderator that influences the meaning of the psychosocial intervention.

13.3   ALIGNMENT OF INTERVENTION WITH CULTURE AND CONTEXT

Psychosocial support should be contextualized in order to reduce ethnocentric attitudes and perspectives in the program for the target communities and to monitor and evaluate design (IASC Reference Group, 2017). Establish congruence between the different groups in the community and within communities in the affected area. The attitude and perspectives can conceal important differences between the home base culture and other cultures that they are compared with.

The American Red Cross developed psychosocial support in several countries during the 2005 South Asia tsunami. Five years later, the programs were evaluated and major differences were reported in the implementation strategies and the ethnocentric attitudes of the participating national societies. A meta-analysis of these evaluation studies was conducted (Powell, 2010). According to Powell, the most important finding was that "psychosocial support was perhaps the most popular project with beneficiaries" (p. 1). As he reviewed the large volume of materials, reports, and monitoring visits, Powell found that "perhaps the most substantial impact of the PSP was the relative unspecific effect that very large numbers of ordinary people in communities, especially women, had involved in a project that brought hope, broadened their horizons, taught them some skills, and increased their feeling of self-efficacy" (p. 41).

The emic results were credited with the success of the program for the affected people. The alignment of intervention with context included: (1) most of the activities under the psychosocial support framework included culture-specific adaptations; (2) use of the program activities to strengthen women's place in the public life of their communities; (3) the project integrated well with religious traditions and encouraged religious activities as a forum for collective celebration and stress reduction; and (4) psychosocial support served as a conduit to continue to address community concerns such as identification of social capital and the reestablishment of place.

Prevalence of cultural imbalance between countries with a large number of expatriate staff yielded different results than countries with most employees being host country native. This was evident in the American Red Cross PSP programs. As time passed and the program evolved, the goals and objectives fluctuated with the influence of mutuality between the affected people and the implementing partner. For example, the initial goal was to "increase well-being," however, in the 5th year of the program the goal was to "reduce psychosocial risks." The meaning of this paradigm shift was that the Red Cross should continue in "etic" environments, focusing on preventive work based on identifying and working with individual and community strengths.

The cross-cultural yield of psychosocial support after the tsunami was, first, that the host country staff would be more appropriate to develop a program than expatriate personnel. The work environment would be different for both groups, and this difference would contribute to the difference in the affected people's values, preference, and attitudes toward the program. Second, the diversity in environmental factors can differ across the affected area and serve as a limitation to develop a psychosocial program. These include topography, religion, economic development, and political boundaries. These issues can be important in PSPs that are developed in multiple countries. Program evaluators and monitoring and evaluation personnel may encounter difficulties assessing differential of familiarity with the program across the culturally diverse groups of responders.

Prewitt Diaz (2008) proposed to align psychosocial support activities with the cultural context of the affected people, and the program staff have to establish cultural equivalence and functional equivalence for the interventions. "Functional equivalence" refers to translating by finding reasonably equivalent words and phrases while following the forms of the source language as closely as possible. When translating a concept, the translator may need to stick close to the form of the source language but use careful paraphrasing to make the translation clear and understandable.

It is a translation method in which the translator attempts to reflect the thought of the writer in the source language rather than the words and forms. The translator will read a sentence or other unit of thought, and ask a representative group of affected people to try to understand it as much as possible, and then write that thought in the target language. The forms of the source language are not important because they are different from the forms of the target language.

Cultural Equivalence: Culture is the lens or template that people use in constructing, defining, and interpreting reality. This definition suggests that people from different cultural contexts and traditions will define and experienced reality in very different ways. Thus, social and psychological reactions to a disaster vary across cultures because they cannot be separated from cultural experience (Marsella, 1982).

Affected people cannot separate their experience of an event from their sensory and linguistic mediation of it. If these differ, the experience across cultures must also differ. If we define who we are in different ways (i.e., self as object), if we process reality in different ways (i.e., self as process), if we define the very nature of what is real, and what is acceptable, and even what is right and wrong, how can we then expect similarities in something as complex as psychosocial reactions to a disaster (Marsella, 1982).

Any drawing or visual stimulus that is delivering a message of empowerment, resilience, distress, and stress emotional reactions must use methods that are responsive to the community where the event has taken place. Different historical and cultural traditions frame disaster experience and the resulting emotional manifestations, within different contexts, thereby promoting and/or shaping different understandings and meanings.

The use of visual messages as tools for psychosocial support takes into consideration the importance of the social and cultural context of preexisting or disaster caused social and psychological problems in understanding the expression of psychosocial reactions to a disaster, and in understanding its assessment, the analysis of information, and designing community interventions that foster resilience. Psychosocial problems emerging from a disaster must be understood within the cultural context that socializes, interprets, and responds to them. This requires that the expatriate personnel and resources proceed from different values, perspectives, and practices, especially those that emphasize context, ecology, and naturalistic methods in the diverse cultural practices of affected people.

Best Practices: Efforts to match interventions and samples from among the affected people is important so that sample difference can be ruled out. This way cultural difference can be ruled out, and the results would be focused on the result of PSP.

1. It is best to identify affected people by age, education, time in community, religion, and work experience, then design a program evaluation based on the interventions. The psychosocial support components (community, sense of place, and schools) may be

categorized the same way, but the activities differ among cultural-linguistic groups and context.

2. Development and administration of materials, educational tools, and instruments should be consistent across context and diverse cultural settings. In order to match interventions and samples, functional and cultural equivalence must be sought prior to the initiation of an intervention and then monitor its results periodically to ascertain that everyone is on track.

13.4 VALIDATION OF MATERIALS, EDUCATIONAL TOOLS, AND INSTRUMENTS

In developing materials, educational tools, and instruments there has to be a basic level equivalence across cultures. In the case of the American Red Cross psychosocial support, the initial language was English, the original translations were made in Hindi and Gujarati. The second generation of the program tools and strategies, during the tsunami, were being translated into Tamil, Divehi, Singhalese, and Bahasa-Indonesia. To establish comparison within and between cultures, two forms of equivalence were utilized: (1) semantic equivalence so that the semantics of the materials and tools could be understood by all language groups participating in the psychosocial program. The method of translation and back translation (Holtzman 1968) was utilized; (2) program implementers ensured that there was conceptual equivalence among the target groups. Related to that is whether the materials and educational tools elicit the same conceptual frame of reference in culturally diverse groups.

13.4.1 SEMANTIC EQUIVALENCE

This section will briefly discuss the methods used to account for linguistic differences among the diverse group of affected people. The discussion is relevant because a parallel between the goal of a program and the translation strategy for materials, training, and monitoring and evaluation has a positive impact on achieving the goal of the program.

A technique of assuring that the contents of material are similar among linguistic diverse groups is the "back translation" method reported by

Brislin (1970, 1986). The objective of this procedure was to develop valid and reliable instruments beginning with the original language of the instrument, translating an instrument from the target language to a second language, translating the instrument from the second language to the target language, resolve linguistic differences in the translation, reach consensus on discrepancies, and develop a finalized version of the instrument in the second language (Banville et al., 2000). Brislin's model was an attempt to maintain equivalence between the original version of the instrument and the translated version.

Cha et al. (2007) modified the back translation approach to ensure content equivalence by dividing the back translation process into four techniques: back translation (Brislin, 1970), the bilingual technique, committee approach, and the pretest procedure. The back translation approach is widely used as a validation tool (Behling and Law, 2000). The bilingual technique recommends the use of independent bilingual translators conducting translation where another bilingual translator will back-translate the instrument (Triandis and Brislin, 1984). This approach has a weakness when compared with the regular translation model. It does not specify how many translators will be needed to validate one instrument.

Another technique is the committee approach. In this approach, a group of bilingual experts translates the instrument from the original language to a second language (Brislin, 1970, 1986). This provides a clearer version since potential mistakes may be corrected among the group of translators. The committee attempted to answer several equivalence questions: (1) semantic equivalence: Do the words mean the same thing? Are there grammatical difficulties in the translation? (2) idiomatic equivalence: colloquialism or idioms that are difficult to translate; (3) experiential equivalence: the items are seeking to capture and experience daily life. Oftentimes, a given task may not be experienced in a different culture; (4) conceptual equivalence: words may have a different conceptual meaning between different cultures (Beaton et al., 2000). Guidelines for the process of cross-cultural adaptation of self-report measures should be strictly adhered too; ad-hoc translation may have an impact on the meaning of the sentence or phase.

The pretest procedure or pilot study is the fourth technique, which requires that once the instrument has been developed in the second language and there is an agreement in the linguistic appropriateness of the items, the instrument is administered with a sample of the target population. This field testing will focus on differences in the response pattern of the sample based on values, preference, and attitudes. Language is contextual and will impact the response in a given sample irrespective of the linguistic appropriateness of the instrument in a second language. The result will yield potential problem that affects equivalence.

Darwish (2003) viewed the translation process in a more holistic manner. He suggests that the most important step in translation is to go beyond the comparison of different textual versions and linguistic systems towards an understanding of how translation operates in totality of all communicative interaction, how communication can take place when different codes are involved, and what the mediating translator does to bring about communication in the target language. He suggests that translation is a continuous decision-making process that is affected by the degree of indeterminacy a source language text might present.

Bolton and Tang (2002) developed a method for assessing the prevalence of specific mental health problems in resource-poor countries and their impact on function. The researchers introduced nonverbal response cards to elicit meaning from an instrument in English to a setting where written language was not used. They were looking specifically at what were the tasks that men and women must do regularly to care for (1) themselves, (2) family, and (3) their community. Based on the results of their research they developed a frequency list and ultimately a list of functions used by the target population in Uganda and Rwanda. Based on the trials they determined was a feasible tool to use (Bolton and Tang, 2002).

In summary, the back translation approach is an excellent tool for seeking and obtaining semantic equivalence. However, in resource-poor countries or locations where written languages do not exist, the committee approach would be appropriate where the translators are knowledgeable of the source language and the language that will be translated and will take time to learn about the context and culture of the target community.

Best Practice: For establishing semantic equivalence the person responsible for implementing PSPs employs back translation during planning, determining the target population and implementing the program with instruments and canned programs developed, and validated in a different language. This will improve the quality of the data. People will reflect their cultural assumptions and values when they are able to communicate feelings in their native language using terms related to culture and context.

1. Instruments and materials should include figures of speech, terminologies, and phrases that are representative of the target communities. The translation process has includes a committee approach with translators from both the original language and the language spoken by the target community. A dual focus will provide more accurate results because it will grasp the culture, context, and language of the sender and the recipient. Constructs and their meaning should apply equally to the different cultures interfacing in a psychosocial support interaction.

2. In cross-cultural programming, the use of short sentences, structured interviews following the administration of an instrument, or teaching of a course will be valuable in facilitating the understanding of content that is culturally and contextually diverse.

13.5 SUMMARY

The purpose of this chapter was to discuss the importance of ascertaining that a cross-cultural interaction between an external stakeholder's rigorous uses and research supported methodology. The chapter discussed three basic points: (1) development of interventions, (2) alignment of culture and context, and (3) validation of materials, educational tools, and instruments. Each section presents best practices. Two case studies are also included that focuses on a transference of cross-cultural methods into a PSP design.

REFERENCES

Banville, D.; Desrosiers, P.; Genet-Volet, Y. Translating Questionnaires and Inventories Using Cross-Cultural Technique. *J. Teach. Phys. Educ.* **2000,** *19,* 374–387.

Beaton, D.; Bombardier, C.; Guilllemin, F.; Bosi-Ferraz, B. Guidelines for the Process of Cross- Cultural Adaptation of Self-Report Measures. *Spine* **2000,** *25*(24), 3186–3191.

Behling, O.; Law, K. S. *Translating Questionnaires and Other Research Instruments: Problems and Solutions;* Sage Publications Inc.: Thousand Oaks, CA, 2000.

Bolton, P.; Tang, A. M. An Alternative Approach to Cross Cultural Function Assessment. *Soc. Psychiatry Epidemiol.* **2002,** *37,* 537–543.

Brislin, R. W. Back Translation for Cross Cultural Research. *J. Cross Cult. Psychol.* **1970,** *1,* 185–216.

Brislin, R. W. The Wording and Translation of Research Instruments. In *Field Methods in Cross Cultural Research;* Lonner, W. L., Berry, J. W., Eds.; Sage Publications, Inc.: Newbury Park, CCA, 1986; pp 137–164.

Cha, E.; Kim, K. H.; Erlen, J. A. Translation of Scales in Cross Cultural Research: Issues and Techniques. *J. Adv. Nurs.* **2007,** *58*(4), 386–395.

Darwish, A. *The Translation Process: A View of the Mind;* Writescope: Melbourne, 2003, http://translocutions.com/translation/mindview.pdf (accessed July 27, 2008).

Dash, S.; Dayal, A.; Lakshminarayana, R. N. The Development of a Linguistically Appropriate Instrument in Two States in India. In *Community Based Psychosocial Support: A Tool for Re-establishing Place;* Prewitt Diaz, J. O., Ed.; Apple Academic Press: Toronto, Canada, 2017.

Holtzman, W. H. Cross-Cultural Studies in Psychology. *Int. J. Psychol.* **1968,** *3*(2), 83–91.

Inter-Agency Standing Committee (IASC). Reference Group for Mental Health and Psychosocial Support in Emergency Settings. A Common Monitoring and Evaluation Framework for Mental Health and Psychosocial Support in Emergency Settings, IASC, Geneva, 2017.

Jordan, J. V. *The Meaning of Mutuality;* Paper presented at Stone Center Colloquium: Boston, Mass, 1986.

Marsella, A. J. Culture and Mental Health: An Overview. In *Cultural Conceptions of Mental Health and Therapy;* Marsella, A. J., White, G., Eds.; G. Reidel/Kluwer: Boston, MA, 1982; pp 359–388.

Powell, S. *Psychosocial Programming: Meta Evaluation American Red Cross*; Palang Merah Indonesia/American Red Cross: Indonesia, 2010.

Prewitt Diaz, J. O. Words into Action: Pictorial Contextualization of the IASC Guidelines on Mental Health and Psychosocial Support in emergency settings. *Intervention* **2008,** *6*(3/4), 327–333.

Prewitt Diaz, J. O.; Lakshminarayana, R.; Bordoloi, S. Psychosocial Support in Events of Mass Destruction: Challenges and Lessons. *Econ. Polit. Wkly.* **2004,** *39,* 2121–2127.

Triandis, H. C. *Culture and Social Behavior*; McGraw-Hill: New York, NY, 1994.

Triandis, H. C.; Brislin, R. W. Cross Cultural Psychology. *Am. Psychol.* **1984,** *39,* 1006–1016.

CHAPTER 14

HURRICANE MARIA, PERSONAL AND COLLECTIVE SUFFERING, AND PSYCHOSOCIAL SUPPORT AS A CROSSCUTTING INTERVENTION

JOSEPH O. PREWITT DIAZ

Chief Executive Officer, Center for Psychosocial Support Solutions, Alexandria, VA 22310, USA

CONTENTS

14.1 INTRODUCTION

Hurricanes and weather events are becoming increasingly common in the Caribbean region, having a significant psychosocial impact on survivors. This impact results in psychosocial distress and can lead to psychosocial disorders and mental health needs. Hurricane Maria made landfall just south of Yabucoa Harbor in Puerto Rico at 6:15 a.m. on September 20, 2017. The National Weather Service recorded maximum sustained winds of 155 miles per hour, making Maria the first Category 4 cyclone to hit the island since 1932 and Maria was just short of being classified as Category 5. Parts of Puerto Rico experienced 30 in. of rain in 1 day, equal to the amount that Houston received over 3 days during Hurricane Harvey. The winds caused tornado-like damage over a swath of the island. The winds were strong enough to destroy the National Weather Service's observing sensors in the territory, forcing meteorologists to measure the storm entirely by satellite.

The storm knocked out the power of the entire island. Much of the island's population were unable to access clean water and were without electrical power. Local officials warned that, in some towns, 80–90% of structures could be destroyed. During the morning of September 21, the day after landfall, rain from the storm continued to deluge Puerto Rico, and the National Weather Service warned of catastrophic flooding in the territory's mountainous interior.

Tailoring psychosocial support to the immediate disaster response might have alleviated emotional suffering to thousands of island residents. Although many of the previous disasters shared many characteristics (such as potentially traumatic events), distinct variations in the type, scope, and population impact of these events and the structure of existing health service systems highlight the need to tailor psychosocial responses.

Most of the island's hospitals were without electricity. The road infrastructure was damaged, and transportation of emergency vehicles, including ambulances, was limited or nonexistent. In some parts of the Central Mountain region, access remained limited for 1 month after the Hurricane. A large portion of the region's mental health providers was affected by the disaster, and would not return to their offices until the curfew was lifted, 3 weeks after the Hurricane. The impact of Hurricane Maria required the mobilization of local, national, and international resources to address the basic needs of the population.

This commentary is a preliminary report on the psychosocial support needs identified during 128 interviews held with survivors during a 10-day period in the immediate aftermath of the hurricane. Here, we attempt to shed light on the comments from PREMA (the Puerto Rico Emergency Management Agency), which suggested that recovery from this disaster had nothing to do with psychosocial support.

14.2 THE HURRICANE

Hurricane Maria made landfall on September 20, 2017 and wreaked havoc on the island causing widespread destruction and disorganization similar to Hurricanes Katrina and Andrew. Meteorologically, Maria was nearly a worst-case scenario for Puerto Rico. The huge center , almost Category 5 hurricane, made a direct hit, lashing the island with wind and rain for more than 30 h, crossing the island from the southeast to the northwest.

The hurricane was catastrophic, destroying a large amount of the built environment, including roads, causing the collapse of the electric grid, shutting down running water for 1 week, and disrupting ground and radio communications. Many elderly and sick patients who depend on electrically-run life support were at extreme risk of losing their lives. A recent report suggests that more than 900 people have died since the hurricane (Osborne, 2017).

It is very difficult to navigate the impact zone and extremely difficult to preposition supplies, many of which might have been destroyed. Thus, Hurricane Maria was more devastating than hurricanes Irma and Harvey, both of which left much of the nearby infrastructure standing. In both storms, supplies that were positioned inland or in Atlanta were available after the storms had passed.

14.3 METHOD

Interviews were held by 10 two-person psychosocial support teams in 22 towns located within the path of the Hurricane. These informal interviews provided a glimpse of the psychosocial impact of the Hurricane. All interviews were conducted within 10 days of the Hurricane (September 22–October 3, 2017).

The respondents were asked three questions: (1) What is your reaction to the hurricane (reactions); (2) Can you describe what you have been doing since the hurricane (risk factors); and (3) What are the next steps (resilience factors)? These factors combined give us a complete view of the person and his/her needs for psychological first aid (PFA) (American Red Cross, 2017). The interviews were recorded and compiled daily in a group meeting. The mean interview time was 30 min. The results are reported below.

One shortcoming of this report is that many people left Puerto Rico in the immediate aftermath of the hurricane, as well as others in the subsequent weeks, and, thus, were unavailable for an interview. We do not yet know how these people will cope in the United States. We have comments from family members who stayed behind:

> *The hurricane separated me from my parents, they have gone to Orlando, which is a strange land, they will never adjust.*

Another respondent had sent her children from the Barrio down to the town to stay with her parents:

> *The girls are gone, they are living in a strange and different place. I feel bad for them. The town is only 2 km from their destroyed house.*

14.4   IMPACT OF HURRICANE MARIA

On the night of September 20, psychosocial distress was felt by all of the island's inhabitants. The extent of the reaction was based on the period that they were exposed to the high winds, the heavy rains, and the distance from vegetation. The feeling of being attacked by the winds, the level of rivers and creeks, and mudslides were frequently reported in the first days after the hurricane. It is as yet unclear how many survivors were psychologically injured as a result of exposure to the hurricane and the subsequent days of deprivation of basic needs, shelter, and water.

As the groups began their visits, they saw people wandering, others in small groups in the street corners, and yet others walking around the neighborhood looking at all the damage and trying to figure out their next step. Most of the interviewees talked about the destruction of the built and natural environment: "Look, everything is broken, covered with mud and destroyed." This was a reflection of the newly opened wounds in the hearts and minds of the survivors. They felt a "hurting" deep inside and did not know what to do, other than to sit and look, and to wait. People were in a state of shock.

Among those who are beginning to recover, there is a sense of loss of family continuity: "the children were sent to live with their aunt in Orlando. This will be the first Christmas away from us; we will not be a family anymore." The other marked comment addresses the loss of houses to mudslides and the wooden shacks due to the wind: "We have collected enough good wood and metal sheets to reconstruct a bigger house." Another respondent talked about the additional space as a result of the destruction. A person in Yabucoa (the town where the hurricane first struck), "now that those three lots are empty, I will be able to plant some *gandules*, banana, and make a vegetable garden."

Meanwhile, cities such as Orlando are dealing with tens of thousands of new immigrants from Puerto Rico 5 weeks after the Hurricane. Many of the reports are from family members who are beginning to develop a sense of vulnerability: "There are too many things happening to us, and all is bad, will we ever recover?" The reported reactions to the hurricane suggest potential negative impacts to psychosocial recovery.

There are 5 million Puerto Ricans, part of a diaspora of the larger cities, in the north and east of the United States. For the millions of people in the Puerto Rican diaspora, the lack of information from and reaching the island is one of the most disruptive parts of the disaster. For people living on the island, it is often impossible to get the word out as conditions deteriorate, to know when aid might be coming, or to coordinate the delivery and access to life-saving services. Their relatives, scattered across the United States and elsewhere, have been left entirely in the dark as to the status of family members (Newkirk, Sept 26, 2017). People are reporting that there is "no safe place to be."

14.5 FINDINGS

For clarity, our findings described below are divided into sections addressing individual distress and loss of community.

The sudden, upsetting experience of living *through* and surviving the *hurricane: What have I done to bring all his misery unto me?*

More than 3 million people survived Hurricane Maria in Puerto Rico. This population experienced major or minor damages caused by winds, rain, and mudslides. One-third of the deaths by drowning occurred in the western part of the island, where a river overflowed. In the Central Mountain Regions, there were severe winds that made the houses fly as though they were kites, mudslides that toppled large concrete houses and buried their occupants in the debris, an urban flood as high as 6 ft that forced the inhabitants to sit on their roofs and watch the destruction around them. It was a horrific site for the survivors.

After nearly 8 h of fighting with the elements, the survivors were depleted of energy, feared for their lives, and were unable to relocate. An older gentleman told us:

> *I tried to go from my roof to my daughter's house next door, the water was up to my chest, the current tried to take me with it, I fought back, I felt that I was being washed away by the current, and finally when I got to the other house (about 15 ft away); I had no energy left, I could not move, and when I realized I could have been washed away I began to shake and cry.*

> *It seemed that the creek was taking its revenge for the years that my family and I had thrown our trash and human waste into the creek, now the waters of the creek were everywhere, the water took two pigs and the chicken house. Inside of my house, about 1 ft came inside through the crack in the doors. The water danced around the rooms as though trying to decide whether it was going to wash us away as well. I see that dance of the water every time I close my eyes and I see and hear the water, as it was deciding whether to take us. It is like a scary dream.*

Their description of events and their immediate responses was as though, as soon as they escaped, they were caught up in fear; their muscles were tensed and then totally relaxed, to the extent that it was difficult to get up, they had no strength and no desire or will to move on.

14.6   I FACED DEATH THAT DAY

The people in the immediate trajectory of Hurricane Maria faced death. The winds exceeded 145 miles per hour; the rain was pouring into the soil that had experienced severe rainfall 2 weeks earlier, when Hurricane Irma barely missed the island, causing unpredictable mudslides in the mountain range that killed and isolated people.

In the town of Utuado, a father thought to protect his three younger daughters; he put them all in a bedroom in the lower level of his concrete home. His wife and older sister remained on the upper level. Sometime in the middle of the night, the soil gave way, and a mudslide buried the three daughters who were sleeping in the lower level (Irizarry Álvarez, 2017). There were human losses all over the island. In the immediate aftermath of the hurricane, there were more than 900 deaths as a result of the loss of electricity, water, and communications. In every town, there were people who had directly or indirectly lost someone to the hurricane.

Such a large number of deaths resulting from a single event do not easily dissipate from our awareness. It remains in our minds and haunts those who witnessed the decomposition and the smell of death.

I was near the house and heard the screams; the creek had taken Dona Margarita, we tried to help her but were not able to take her out. Two days later we found her body by the bridge on the road. It smelled very bad, but we had to take her out so that we could do a Christian burial.

There were over 130 cadavers in the morgue; the refrigeration was not working very well, you could feel the smell of death three blocks away.

14.7 WHY ME?

The interviewees repeated similar comments regarding the survival of some and the death of others. About a woman who was carried away by the creek with her sister: "I do not know why my sister died and I am still alive." A son had just gotten out of the car to remove some debris from the road, and a mudslide covered him. His father and mother were in the car. "Why did God take my son and not me? I am ready, he was young, strong, and had so much to contribute." There is a general feeling expressed by many on the island that the survivors have lived for a reason: "Maybe I am alive for a reason." This was often attributed to divine intervention. Also, as a result of survivor guilt, people have begun to isolate themselves.

Blame for what happens is another source of stress. I am guilty because…. "I allowed him to come and pick me up," "I was not strong and agile enough to hold on to my sister," "This all happened because of the Hurricane," or "If we had moved to the States last year this would not have happened." We expect that guilt and blame will accompany the survivors for a long time.

14.8   I LOST EVERYTHING THAT DEFINED WHO I WAS

The people of Puerto Rico have a tradition of saving mementos of their life (e.g., their grandmother's rocking chairs, photos, baby clothes, rosary beads, diplomas, medals, and books). Every home contains articles that are used as conversation pieces, detailing who the inhabitants are, where they have been, and how they are related to the community or national events. The loss of those possessions may have a material value, but in terms of memorializing one's family's life, they are invaluable. The loss of a home is more than the loss of a structure; it is the loss of the collective history of a family, dating back several generations.

Survivors express how their homes were a status symbol; they are,

I had two albums and some pictures on thru wall they are gone, the loss of the place we know as home is irreplaceable and will be forever part of our loss. I will mourn my home forever.

You do not know the depth of my emotional involvement in building and maintaining my home, and careful crafting of the legacy I wanted to leave to my children and grandchildren.

We are alive, but what defines us is gone forever.

14.9   I HAVE TO GET AWAY FROM HERE—THE PUSH AND PULL FACTOR

The loss of order, everyday routine, and the familiar environment affected many Hurricane Maria survivors: "I have lost my secure space; I have to get out of here." "The government did not do enough to protect us; the economic situation of the island is the blame for all that has happened to us; how could this happen to us?" On the other hand, as soon as the immediate disaster recovery was underway, many people were called to their family's home in the United States. In the 30 days after the Hurricane, over 50,000 people left the island. This includes the elderly, young professionals, university students, and other assorted technical employees. This sudden move is an example of a psychosocial response to a catastrophic event, resulting in a Puerto Rican brain drain.

14.10   SENSE OF LOSS OF OUR COMMUNITY

Following Hurricane Maria, the affected people have been left alone, their natural and built environment have been destroyed, and they feel alone and psychosocially vulnerable. The barrio has been destroyed, the rooftops are gone, and the trees and other vegetation are no longer able to provide a layer of insulation from the outside dangers. Their neighbors have also been deeply affected or have left the immediacy of the barrio. There is no one available to warn you of imminent dangers, no one to care for you if you have an emergency or sudden illness. If you have an illness that requires electricity to run medical devices, there is no one around to take you to the hospital. And even if there was, the roads might be blocked, or the hospital may not be working, or have the necessary life support equipment that you need.

The survivors have to deal with the visual images of the destruction without the filter of someone close (e.g., a neighbor to help filter the emotional impact of those images). How many survivors are paralyzed by imaginary fears of being alone on this emotional voyage? Years of living in a community, in a specific geographic space, and a particular environment, generates a culture unique to that community. In the immediate post-disaster interval, culture serves as a filter, thereby giving meaning to psychosocial fears and needs. Culture helps the survivor to make sense of a precarious situation and to reduce the anxieties that one might feel while negotiating through these radical changes. Hurricane Maria took away the perceptions of being supported by the community; it stripped the illusion of having a place.

14.11   SENSE OF DISORIENTATION

Most people interviewed during the first week after the Hurricane felt disoriented. The world as they knew it had been destroyed. They had lost their community and emotional anchors to their neighbors and significant others in the community.

Some examples of responses are as follows:

Every morning I leave the shelter and walk to my neighborhood. Sometimes I cannot find my way, and other times I just look around and start crying.

I cannot recognize my neighborhood, the houses are gone, so much empty space and the people are not around.

In the neighborhood, everyone walks around with nowhere to go.

I am staying with my daughter in Yabucoa. I wanted to come and see my neighbors, but when I got there, all the houses were gone. When I did not see my neighbor's house I became very nervous and started crying, my chest was tight, and I could not breathe. I thought they were gone forever. I sat in a tree trunk and looked around; nothing fit the image of my "barrio" in my mind.

My daughter stayed with me for a while and then took me back to her house. I was very nervous and did not know what to do or to think; 1 day I had a beautiful little house, and wonderful neighbors, plants, and trees all around. The next day everything was gone, the land was barren of trees, and my neighbors were gone. I do not know where I am or where I am going.

All is a desolation; I do not know where I am right now. The wind came and took everything. Everything. Even my soul.

14.12   SENSE OF LOSS OF CONNECTION

The hurricane disconnected the family, the neighborhood, and the community. Many of the respondents felt torn about leaving to go somewhere else. They expressed their sorrow, anger, and disbelief. It was as if the island had not protected them at the moment of destruction. A man looked at the El Yunque mountain range in the distance and at the scars of discoloration of the soil left by the fast winds: "You, my mountain, could not protect us, but you took a heavy blow protecting the ones on the other side. I am not sure if I will ever see your greenness again."

It is time to move somewhere else; I feel lonely. My family left for Florida, but I will never leave the island. God! I am sad; I hear all that has happened. Give me peace.

I yearn for the sound of the coqui's (Puerto Rican singing frogs), and the smell of the floors. Nature has been destroyed forever.

The people of the neighborhood had been left to themselves. They had an image of themselves that is no more. Even as they walked the streets of the neighboring town, they did not feel the same; nobody could tell who they were; they were different, or so they perceived themselves.

It looked like the survivors were unable to reconcile who they are with who they had been before the hurricane. They feel incapable of relating to others from a different part of town because they perceived them as different.

14.13 SENSE OF VULNERABILITY

Many of the respondents expressed fear of being in their familiar places (their neighborhood and town). It is like their immunity to the bad things in the environment and the external forces of nature has been lost.

Puerto Rico is no longer a good place to live.

There is no electricity, water, the roads are blocked, and no communication. We are desperate. Things are going to get worse, and we are trapped here.

The survivors' friends and neighbors have left, and there is nobody around to trust. Many people thought that by living in the mountain towns they were protected from bad things happening on the island; nothing bad could happen to them, they were tough and had survived many things. It was like the mountains would protect the people in the valleys.

Following Hurricane Maria, fatalism could be identified as a personality trait among the island natives. Now, it seems that the people have lost the capacity to screen signs of danger. Every external event, even the arrival of a vehicle with much-needed aid, is welcomed with great suspicion.

Aid workers reported that, as their truck was traveling on a side road down a mountain range, they were confronted by machete-wielding men, who asked: "What do you want from us?" It took time to convince them that the truck was not there to take away things but to share much-needed water and canned food.

Things have gone from bad to worse; as time passes and people have to resort to primitive ways of caring for each other, their capacity to screen out the events around them (good and bad) is no longer in their awareness.

> *I cannot be still if I go over there, I want to come back here (si me voy pa'ya quiero regresar pa'ca). I am waiting for something even worse to happen, and I will lose my life.*

Fear can be measured by the tired faces of the elderly who claim that they cannot sleep, an increase in misunderstanding, cursing, and physical confrontations, and the use of alcohol. Everyone has a story about what happened on September 20th. Many of those stories are about the irreplaceable loss of natural place and the people who have left the place never to come back. Puerto Rico is a religious country, but people have begun expressing doubts about God's intentions. This is evidenced by expressions of helpless and hopelessness:

> *I do not want to do anything; I am going to stay here (in the ruins of a house), I do not even want to pray, sing praise songs or go to church. God did not listen, I cannot even trust Him anymore.*

14.14   PSYCHOSOCIAL INTERVENTIONS IN THE IMMEDIATE AFTERMATH

14.14.1   PSYCHOLOGICAL FIRST AID

Psychological first aid (PFA) is a first order intervention that promotes calmness and facilitates connection. Psychological first aid is conceptualized as documenting and operationalizing good common sense; those activities that sensible, caring human beings would do for each other anyway (Forbes et al., 2011). In simple terms, PFA includes the provision of information, comfort, emotional care, and instrumental support to those exposed to an extreme event, with assistance provided in a stepwise fashion tailored to the person's needs (Forbes et al., 2011).

In this initial response, the American Red Cross model was utilized (PFA: Helping Others in Times of Stress, 2017). The objective of the intervention was fourfold: (1) to create a compassionate environment for disaster survivors and workers; (2) help a person identify what he/she might need; (3) provide immediate support to those in stressful situations; and (4) help others cope in the face of stressful events.

14.14.2 ASSESSMENT AND LINKING

In the immediate aftermath of Hurricane Maria, people were disoriented and needed to compose themselves. There were two activities that the responders found that worked. At an individual level, people needed others to know that they were alive. They wanted to hear that dear ones knew about them. An example of this activity was an international NGO, which provided satellite phones and satellite uplinks so that families in the affected areas could link with families in the United States. This activity resulted in a feeling of connection with others and the value of knowing that others were thinking and acting on their behalf.

At the community level, people had to feel empowered, to think, and act as though they were active participants in their recovery. This exercise reduced the initial stress and fear of being alone. One example of this action occurred in the town of Cayey (Peres Arroyo, 2017). The mayor of the town realized that this event was too big for him and his employees to handle. The third day after the hurricane he set out to identify a "leader" for every street in the municipality. These street leaders mapped their respective streets, identified the damage, prioritized the needs, and engaged the inhabitants in a civic campaign to clean up. Also, the leaders, with the civic engagements of the neighbors, facilitated the distribution of at least one hot meal a day and water.

In the Command Post there was a map of the "barrios" (communities), and every day there were updates of the needs of the people. The street leaders were able to find and take back medicines and needed insulin for many of the elderly, as well as food and water. They also took back valuable information about how other communities were holding up, and the recovery plan for the municipality. This activity changed the paradigm from hopelessness and stagnation to moving the people into action. This model instilled hope to the survivors.

14.14.3 RECOMMENDATION FOR INSERTION OF PSYCHOSOCIAL SUPPORT FOLLOWING FUTURE CATASTROPHIC EVENTS

The impact of Hurricane Maria in Puerto Rico resulted in the destruction of the infrastructure, and a significant loss of life. The level of previous disaster exposure and preparedness varied considerably in the past 50 years and gave disaster planners at the Federal Emergency Management Agency and the local disaster management agency a false sense of their capacity to manage the impending disaster. Contingency planning was inadequate at best. From my point of view, tailoring psychosocial disaster responses to this specific disaster would have alleviated the emotional suffering of thousands of island residents. While many of the previous disasters shared many characteristics (such as potentially traumatic events), distinct variations in the type, scope, and population impact of these events and the structure of existing health service systems highlighted the need to tailor psychosocial responses.

For example, hospitals were without light across most of the island. The road infrastructure was damaged, and transportation of emergency vehicles, including ambulances, was limited or nonexistent. In some locations of the Central Mountain region, access remained limited for 1 month after the Hurricane. A large portion of mental health providers were survivors and would not return to their offices until the curfew was lifted, 3 weeks after the hurricane. The impact of Hurricane Maria required the mobilization of local, national, and international resources to address the basic needs of the population.

14.14.3.1 TAILOR THE MENTAL HEALTH AND PSYCHOSOCIAL SUPPORT RESPONSE TO THE DISASTER

It will take some time to fully implement a mental health and psychosocial support program that addresses issues of traumatic stress, as well as community's social and emotional needs. These services, when implemented, will have to be tailored to the geographic areas, the perceived extent of the loss, the affected population, preexisting mental health services, and cultural factors associated with mental health and psychosocial support. These are all predictors of post- and peritraumatic stress. Some natural disasters that destroy large parts of the infrastructure and lead to the break-up and dispersion of communities are also accompanied by substantial mental health impacts (Galea et al., 2007). There are a variety of risk and resilience factors that need to be considered when planning the psychosocial response to disasters such as Hurricane Maria (Bonanno et al., 2010). For example, the American Red Cross provides practical assistance, including linking, first aid, and PFA. By the second week after the hurricane, the response included psychoeducation, and crisis and grief counseling to family members and first responders, as well as to the wider community. In the coming weeks, as local facilities begin to operate, the Red Cross will be making referrals for those in need of long-term care.

14.14.3.2   TARGETING AT-RISK POPULATIONS

The identification and targeting of support to particular at-risk groups constituted a key task for disaster responses. All responses sought to target direct disaster survivors and other at-risk groups. Population-based psychosocial support requires assessment and health surveillance of high-risk groups and is of immense practical importance to estimating psychosocial service needs and targeting interventions following a disaster. It is important that we draw lessons from earlier catastrophes and integrate them in the service delivery to affected populations.

Barriers to access to care (providing basic needs and psychosocial support) have been identified in the Central, Southeast, and Southwest parts of Puerto Rico. These barriers include aspects associated with the disaster response, such as the identification of survivors, limitations of existing referral pathways, destroyed infrastructure, and other difficulties in accessing appropriate care. Screening provides a mechanism to identify survivors in need of treatment and to target interventions. For example, the American Red Cross has provided PFA and psycho-education (including information on disaster mental health care, self-care, and available services). Future attempts to improve service accessibility might need to take into account other factors (such as gender, age, disability, socioeconomic status, language, or culture) that can impact on access to disaster care.

14.14.3.3 RECOGNIZING THE SOCIAL DIMENSIONS AND SOURCES OF RESILIENCE

Large hurricane responses in Puerto Rico, such as those to Hugo (in 1989) and Andrew (in 1992), highlighted the efforts to promote a positive recovery environment that is based on the knowledge of the social, language, and contextual dimensions of resilience and recovery within the context of Puerto Rico. Planners and responders need to acknowledge the importance of the Puerto Rican diaspora to the mainland and their immediate response to the affected population on the island. They also have to acknowledge that recovery is different for residents of the urban metro area than for the millions who live in small towns and the central mountain regions.

From the psychosocial perspective, it is important to understand the place of community-level and self-help initiatives that need to be recognized and fostered within disaster response planning, alongside more formal psychosocial support strategies. Social support and bonding are also important to reduce negative psychosocial outcomes after trauma. One of the key tasks for disaster response planners, therefore, consists of recognizing both the value of existing and emerging support networks of those affected by disaster (as well as their limitations) within a broader framework of psychosocial disaster care. In this context, community- and family-based supports and targeted capacity building initiatives deserve particular consideration (SPHERE Project 2012; Inter-Agency Standing Committee, 2007).

REFERENCE

American Red Cross Model was Utilized "Psychological First Aid: Helping Others in Times of Stress". Washington, D.C., 2017.

Bonanno, G. A.; Brewin, C. R.; Kaniasty, K.; LaGreca, A. M. Weighing the Costs of Disaster: Consequences, Risks, and Resilience in Individuals, Families, and Communities. *Psychol. Sci. Public Interest.* **2010,** *11,* 1–49.

Forbes, D.; Lewis, V.; Varker, T.; Phelps, A.; O'Donnell, M.; Wade, D. J.; Ruzek, J. I.; Watson, P.; Bryant, R. A.; Creamer, M. Psychological First Aid Following Trauma: Implementation and Evaluation Framework for High-Risk Organizations. *Psychiatry* **2011,** *74,* 224–239.

Galea, S.; Brewin, C. R.; Gruber, M.; Jones, R. T.; King, D.; King, L.; et al. Exposure to Hurricane-Related Stressors and Mental Illness After Hurricane Katrina. *Arch. Gen. Psychiatry* **2007,** *64,* 1427–1434.

Gurwitch, R.; Hughes, L.; Porter, B.; Schreiber, M.; Bagwell Kukor, M.; Herrmann, J.; Yin, R. *Coping in Today's World: Psychological First Aid and Resilience for Families, Friends, and Neighbors: Instructor's Manual*; American Red Cross: Washington DC, 2010a.

Gurwitch, R.; Hughes, L.; Porter, B.; Schreiber, M.; Bagwell Kukor, M.; Herrmann, J.; Yin, R. *Coping in Today's World: Psychological First Aid and Resilience for Families, Friends, and Neighbors: Participant's Manual*; American Red Cross: Washington DC, 2010b.

Irizarry Álvarez, F. Mueren Tres semanas por derrumbe en Utuado tras paso de María. San Juan, PR: *Diario Primera Hora*, Sept. 21, 2017, http://www.primerahora.com/noticias/puerto-rico/nota/muerentreshermanasporderrumbeenutuadotraspasodemaria-1246727/.

Newkirk, V. R. How Puerto Ricans on the Mainland Are Getting News from Relatives. *The Atlantic Magazine*. Sept 26, 2017, https://www.theatlantic.com/politics/archive/2017/09/how-puerto-ricans-on-the-mainland-are-getting-news-from-relatives/541062/.

Osborne, S. Hurricane Maria: More than 900 People in Puerto Rico Have Died After the Tropical Storm. *Independent News.*: The United Kingdom, 30 October 2017.

Pares Arroyo, M. Un ejercito de lideres se une para levanter a Cayey. San Juan, PR: *El Nuevo Dia*. Revista Dominical. October 29, 2017.

Shultz, J. M.; David Forbes, D. Psychological First Aid: Rapid Proliferation and the Search for Evidence. *Disaster Health* **2014,** *2,* 3–12.

CHAPTER 15

PLACE-BASED PSYCHOSOCIAL INTERVENTIONS IN THE AFTERMATH OF HURRICANE MARIA IN PUERTO RICO[1]

JOSEPH O. PREWITT DIAZ

Chief Executive Officer, Center for Psychosocial Support Solutions, Alexandria, VA 22310, USA

CONTENTS

---

[1] The author would like to acknowledge the review and comments made by Ms. Anjana Dayal. She is currently in Puerto Rico working with community groups to define their recovery. Her insights were extremely valuable for this article.

Originally published in WebmedCentral, October 2017. Reprinted with permission. https://creativecommons.org/licenses/by/3.0/

15.1 INTRODUCTION

The objectives of this chapter are to present (1) a history of Puerto Rico with respect to past hurricanes, (2) a literature review of place-based interventions, (3) a practical place-based psychosocial support strategy to assist affected communities in the recovery and reestablishment of place, and (4) a summary. The chapter concludes that place-based interventions are nimble, cost-effective, community-owned, and bring communities together after a disaster.

Hurricane Maria has caused the inhabitants of Puerto Rico to lose their "sense of place." The place can be described as a geographical, physical, and psychological space where a person belongs, the relation between persons and the natural and built environment, and the interaction over time between both. Humans and nature share a common space. The bonds are so strong that if something happens within the space, the perception of security, intimacy, and well-being are violated, this generates a physical and emotional response from the people.

Place-based intervention includes the affected people in defining what they need to feel better, participate in the planning of "reestablishment of place," and becoming active actors in the remaking of "place." While the process may be similar and fit into a cookie cutter pattern, the internal dynamic of the process is different for every community. Thus, place-based interventions are most appropriate in settings such as the recovery of Puerto Rico after Hurricane Maria.

15.2 BACKGROUND

Puerto Rico is a small island in the Caribbean, the lesser of the major Antilles. The island is approximately 100 miles long and 35 miles wide. Its topography consists of mountain ranges, beautiful valleys, and beaches. The south-western part of the island is arid. Politically, the island is a protectorate of the United States and its citizens have American citizenship. They are taxed but have no representation in Congress.

The island is affected by moderate-to-severe storms every year. The people who experience these crises have developed a hurricane response toolbox that helps repair homes and living environments so the population can move forward. In its history, it has experienced large hurricanes that have devastated the island, its natural beauty, and its people. Over the years, construction has made the island's housing and infrastructure more resilient to these severe weather events. Housing mostly consists of concrete with crossed-iron reinforcements on the walls. Most people in the inland towns and mountain regions own their own homes. The infrastructure in terms of water, electricity, and the road system on most of the island is poor to moderate. As a result, Puerto Ricans have learned to adjust, and, when necessary, have modified their lifestyle to meet the challenges of the climate.

15.2.1 CULTURE

Puerto Ricans love the flora and fauna of the island. The tropical weather is pleasant, the people are caring, and they judge their life based on the natural beauty of the island. Their culture is paramount and it sets the stage for recovery. Most Puerto Ricans dedicate popular music, poetry, and other forms of artistic expression to the island (Puerto Rico Patria Mia, Verde Luz, Preciosa). Even in the initial days after the hurricane, neighbors, and families were meeting together for collective cookouts, singing *Plenas* (typical music), and playing traditional instruments (la Pandora). It was impressive to see the broken environment and hundreds of cars on the highway looking for connectivity. Many of those were just looking for someone to talk to, dealing with their loneliness and fears.

A cultural sense of place is denoted by various activities, such as community gatherings where information is exchanged, groups learning about customs and traditions, and discussions focused on learning from a common history. These gatherings can be around faith, music, meals, education, or other local ways of coming together to share information and support each other. The culture typically takes the longest time to reestablish a "sense of place" in the community because it involves recovering history.

15.2.2 ENVIRONMENT

Puerto Rico—the island and its people—were damaged by the ferocious winds and rains of Hurricane Maria. This was the most powerful hurricane to affect the island in more than 100 years. Most of the concrete houses survived, the trees were torn from their roots, and the infrastructure was so damaged that it will take months for the island to regain electricity, clean water, communications, adequate medical care, and medicine. Some concrete structures were lost to mudslides and overflowed rivers. When people saw the vegetation destroyed, they were emotionally impacted. The general sentiment was "if the island is destroyed, what will happen to me?" By the 3rd week after the hurricane, the island began to provide positive reinforcement to the people who love her. The trees began to show new leaves even though many of the roots were still exposed. The flowers began to grow in the gardens, and the tropical sweet smells of fresh vegetation were present all around. While most people are still without electricity, water, or jobs they are celebrating that the mountains are turning green again, the trees are growing leaves, and the flowers have begun to blossom.

Environmental activities related to reestablish a sense of place include the community's knowledge of, and involvement with native plants and wildlife, the extent to which the natural environment was destroyed by the disaster, community plans to preserve natural resources for future generations, and the extent to which community members actively participate in environmental conservation and restoration activities. Helping people reestablish their ecological sense of place involves identifying ecological segments of the community that were destroyed by the disaster, utilizing the community's knowledge of native plants and wildlife, exploring and explaining the interactions between individuals and the environment, establishing a level of comfort between individuals and the environment, and then bringing those ecological segments back into the community. In the case of Puerto Rico, a tropical island, it is incredible to see how most of the vegetation is quickly coming back to life on its own. In less than a month, there are places which are already green. However, the impact on the environment is still very visible. For example, in the eastern part of the island, the palm trees are standing but the branches have been turned upside down. The reactions of the local people are expressions of pain, similar to when a child in the community is injured.

15.2.3   SOLUTION FOCUSED

People were so distressed that in the first 2 weeks, over 30,000 relocated to Miami. Many others began to use their transformational power to figure out what they needed and determine how to change the needs into opportunities to improvise and succeed. For example, there are bankers, teachers, and other professionals serving as drivers, interpreters, and guides for various governmental and nongovernmental organizations that have arrived at the island to support the recovery. Their efforts are a testimony to the positive effects of coping, which was exhibited by people who were surviving, facing a new reality, and heading back to work in spite of very difficult circumstances.

A solution focus on place-based activity to reestablish the "sense of place" requires critical thinking and a genuine dialogue for community-led answers to the problems caused by the hurricane, rather than focusing on a list of items lost or how the hurricane exacerbated existing problems. A community must develop long-term solutions by nurturing whatever resources and strengths it possesses, and by making them sustainable on both individual and community levels. Ultimately, disaster-affected communities must be the architects of their own psychosocial recovery and what the reestablished place will look like.

15.2.4 SOCIAL CAPITAL

Puerto Rico has always enjoyed the benefits of what is called the "underground economy." The Puerto Rican people use their talents to help each other recover. Some examples presented in the media are:

- An island mayor drives to an underpass to communicate and sends runners with messages so that city business is conducted. This man is doing the duty he was elected to do.
- Stores and restaurants are opening and operating outdoors.
- Since there is no electricity and cell phone service, people have begun the old tradition of hanging out in small groups and talking about solving existing community problems.
- People are getting water for drinking and bathing from sources in the hills. Trucks collect the water and sell it in neighborhoods to people who do not have the wherewithal.
- Many people are scrounging for materials that were displaced by the hurricane so they can rebuild or sell these items in the metro area.

- People that have electric generators are inviting the neighbors to share a meal, have some ice, and charge their mobile phones.

Neighbors helping neighbors may be a positive tool for reconstructing the community. This gives people ownership over community interventions and fosters a sense of resiliency in the affected communities. At the same time, this provides an infusion of financial resources by implementing programs, such as cash-for-work. Potential resources may include a retired carpenter willing to train younger members of the community, some unused land that could be used for a communal facility, unemployed youth who can provide energy and enthusiasm, a farmer or other food producer and some people willing to prepare food for communal laborers donating their time and energy, and/or some trustworthy community members willing to put in time and effort to design a community project.

15.3   LITERATURE REVIEW ON PLACE-BASED INTERVENTIONS

15.3.1   PSYCHOSOCIAL IMPACT OF HURRICANE MARIA

Disasters, whether natural or human-induced, damage the location that they impact and the emotional fabric of the survivors. Physical wounds, which can be seen, can be treated and will heal when given sufficient time. Emotional wounds, which are not seen, exist in the psyche and the memories of those affected and may linger because they go unnoticed.

Hurricane Maria instantly fractured the social and psychological community networks as evidenced by the displacement and relocation of thousands of individuals to Florida in the first 2 weeks after the hurricane. Thousands of people lost their sense of place. When the physical, social, and psychological place is destroyed, disaster-affected people grieve their place in ways similar to mourning a death. The loss of access to places of cultural and social significance and the resulting loss of connections to other affected people undermine a community's ability to act. This can exacerbate the grief.

Disaster-affected people often feel out of control and experience helplessness and vulnerability. Hurricane Maria disrupted and destroyed the lifestyles, places, and feelings of safety of the people throughout the island of Puerto Rico. The impact of losing access to running water, electricity, and communication can cause people to see the world as threatening. This results in some individuals banding together in groups to help reestablish boundaries and structures, thereby creating a new sense of place. That is the crisis brought about by the impact of Hurricane Maria, which causes a sense of loss and the desire to escape from what is now perceived as a threat. It also fosters creative problem-solving and a positive transformation from the chaos that now exists. Even when the community reacts in a productive manner to the impact of the hurricane, changes still occur with respect to the social and emotional lives of individuals, the resilience of families, and the cultural fabric of communities.

Hurricane Maria was an extraordinary event that occurred across the island. In one night, it contributed to the collapse of the existing individual, familial, and societal structures. That caused considerable harm to the physical, social, and psychological environment. The results of this event will be felt for many years to come.

The impact of Hurricane Maria and the combined effects related to the damage to the natural environment and the destruction of the state's infrastructure has negatively impacted the place and spirit of the people. Moreover, the reliability of social networks, the government, and private community support systems has proven ineffective in alleviating suffering.

15.3.2 IMPORTANCE OF PLACE FOR COMMUNITY MEMBERS

Fullilove (1996) defined "place" as a setting in which people feel that they have a sufficient living environment. The perception of disaster-affected people is linked to the surrounding environment through three key psychological processes: attachment, familiarity, and identity. Place attachment involves a mutual caretaking bond between a person and a beloved place. Familiarity refers to the processes by which people develop detailed cognitive knowledge of their environment. Place identity is concerned with the extraction of a sense of self, based on the places in which an individual spends his or her life. Each of these three psychological processes can be threatened by displacement; consequently, problems associated with nostalgia, disorientation, and alienation may ensue (Fullilove, 1996, p. 1516).

Williams and Stewart (1998) functionally defined "place" as a collection of meanings, beliefs, symbols, values, and feelings that individuals or groups associate with a specific locality. A disruption of place impairs a person's and community's ability to integrate his or her past with his or her present life because of the lack of tangible social and environmental cues and symbols. This disruption of place may be manifested in the fracture of emotional bonds that people form over time; the loss of strongly held values, meanings, and symbols; the loss of the quality of place, which may be taken for granted until it is threatened or lost; the loss of place meanings that are actively and consciously constructed and reconstructed within affected peoples' minds; and/or the loss of shared cultures and social practices (Fullilove, 1996).

Loss of place has a negative influence on the physical, social, and psychological well-being of disaster-affected people. Shalev and Ursano (2003) indicated that physical needs, such as hunger, pain, or dehydration, can cause people to feel insecure and apprehensive about their future.

Furthermore, Davis et al. (2010) found that the loss of familiarity with pre-disaster social networks, community structures, and financial and personal resources disrupts social, communal, and regular daily living. This generates significant distress. This has been the case in Puerto Rico where long lines at gas stations, the loss of electricity and water, the inability to secure funds at ATM machines, and the disruptions in daily routines are blamed by the local press for the distress expressed by people, in addition to the increase in emergency room visits for physical and psychological symptoms.

More than 30,000 people have left the island in the last 2 weeks. The press reports that 100,000 people will relocate permanently to the United States. While disaster-affected people may feel safer and happier in this new setting, they, too, will be emotionally affected as time passes because of the impact of the new environment. The forced separation from Puerto Rico, as a result of Hurricane Maria, uproots people from their cherished neighborhoods and communities, familiar environments, and from the certainties of life. The uncertainty related to when the recovery will be completed and they can return to Puerto Rico can cause extreme distress, enhancing feelings of separation, disconnection, detachment, and the inability to relate with others (McFarlane and Williams, 2012; Norris, 2002; Shalev and Ursano, 2003). Clearly, programs aimed at helping disaster-affected people overcome these problems associated with the disaster, and loss of place would be beneficial.

15.3.3 PSYCHOSOCIAL SUPPORT AS A CONDUIT FOR PLACE-BASED INTERVENTIONS

Hurricane Maria offers the opportunity to develop place-based interventions that will reconnect families and neighbors. Psychosocial support is an accepted practice during the recovery and reconstruction phases following natural disasters. Saraceno (2006) suggests that psychosocial support ameliorates the negative reactions to enormous losses, such as grief, displacement, disorientation, and alienation. These effects are often ignored in the immediate aftermath of natural disasters, and forgotten during the reconstruction phase. Psychosocial support builds on the knowledge, awareness of local needs, and protective factors to provide psychological as well as social support to people affected by natural disasters. The aims of place-based psychosocial support activities are to enhance the resilience of disaster-affected people to achieve psychological competence by empowering them to overcome grief and move forward in a collaborative fashion.

There are several steps that need to be taken with the community as the primary actor: (1) input should be obtained from all community members through community mapping exercises, (2) systematic information should be gathered to assist the community in prioritizing its perceived needs, (3) community resources and human capital should be identified, and (4) community members should be involved in planning, developing, monitoring, and reporting the recovery projects (Prewitt Diaz and Dayal, 2007).

Psychosocial support reestablishes the social and psychological well-being of people and it provides the tools that individuals can use to rebuild their social networks. This approach provides the tools and the space so that people recognize the emotional impacts of the disaster, express their feelings, and initiate a process of reconstructing their lives within the social networks in their neighborhoods and communities.

Places may act as thresholds through which the living can contact those who have gone before, with human and nonhuman lives born and yet to come. If one takes root shock seriously, then caring for a place is a way to repair our worlds to help create socially sustainable cities across generations. Through place-based psychosocial support activities, affected people are called upon to be resident experts and perceive the disaster-destroyed location as an inhabited place. In this way, affected people slowly begin to remember and there are possibilities that shared belongings can occur in a reestablished place. Through psychosocial support activities, affected people accept responsibility for being unprepared for a disaster, engage in mourning activities, begin a process of emotional transformation, and, as a group, visualize and develop a new place. Addressing loss becomes a dynamic developmental path in which we discard the ruins caused by disasters, and embrace a new beginning in a reestablished place.

Place-based psychosocial support recognizes that affected people themselves are the subjects of their own recovery by recognizing their personal emotional needs, the community environment, and the steps that should be taken for reconstruction. It is an integral, interdisciplinary, inter-institutional, and intersectoral process (Prewitt Diaz, 2008) that allows the individuals, families, and communities to reestablish their resiliency and their social, functional, and psychological development in such a way that they can resume or recreate their life.

15.4 PLACE-BASED PSYCHOSOCIAL SUPPORT STRATEGY

Several factors interact with respect to the growth and welfare of an affected person. These factors are geographical (space and ecological factors), physical (food, shelter, protection, and medical care), psychological, (attachment, affection, and self-esteem), spiritual (belief system, identity, and values), and social (family, friends, and a sense of place to which one belongs). A person belongs to and interacts with the community in which he/she find himself/herself, regardless of whether that community is original or adopted. Therefore, to enhance recovery, reconstruction, and resilience, we should encourage affected people to act at the individual, family, place, and community levels.

Prewitt Diaz (2013) suggested that some place-based psychosocial support activities that foster place attachment include:

1. Enhancing the disaster-affected peoples' capacity to prepare their own place development plan with strategies for implementation;
2. Increase the information base by mapping the place to consider natural systems, the built environment, and psychosocial systems in the area while also identifying the available social and environmental capital;

3. Facilitating the community's decision-making capacity by encouraging the participation of all community segments;
4. Enhancing negotiations and conflict resolution capacities for all disaster-affected people involved in the process;
5. Facilitating involvement of other outside agencies or groups as needed to improve the psychosocial, cultural, ecological, economic, and legal environment in the target place within the larger community; and
6. Designing and executing adaptive resilience projects.

Place-based psychosocial support empowers neighborhoods and communities to identify their cultural, ecological, and social capital to improve collective efficacy and alleviate fear in children and adults alike during the reconstruction phase. This approach is a compassionate tool that provides support and equal access to services for all segments of the community. Place-based activities:

• Are participatory in nature and provide a mechanism whereby all segments of a community can identify the risk and resilience factors in their geographic, ecological, cultural, economic, spiritual, social, and psychological place.
• Provide a physical and temporal space for all members of a community to identify their losses, what they need to rebuild, their social capital, and what the affected communities need from outsiders and other stakeholders.

A disaster emotionally transforms affected people (Hobfoll et al. 2008). To understand place meaning, one must understand emotional transformations of space to place. The emotional energy while sharing stories will lead the community to explore and develop strategies to move forward in reestablishing place. In a community, the participants of a meeting may experience two types of emotions: feelings of the experiences and feelings of sharing them with others in a public forum. Feelings of the lived experiences immediately associate the affected person with his or her environment in ways that can clearly be understood by the audience.

The role of place-based psychosocial support in its initial stages is to seek place meaning, regardless of where the stakeholders believe they are in the disaster cycle. These activities promote an understanding of the sense of place predicated by the changing experiences, feelings, and memories of the affected people. How affected people feel about their neighbors, families, communities, and places of the past and present will shape their memories and their stories into an emerging new place.

Place-based activities provide a space for reestablishing a sense of place. The sense of place refers to psychosocial support activities that help people face the impact of surviving personal losses after Hurricane Maria. Survivors examine the ways they think and act in order to reconstruct their lives as environmental, social, and ecological changes take place during the reconstruction phase. Place-based psychosocial support activities identify survivors as the primary actors in the reestablishment of a sense of place. Representatives of all sectors and groups from the community are actively engaged in making communal decisions, making the time and effort to choose their goals, identifying resources, and formulating their action plans. All of these actions help empower them and their communities.

Reestablishing a sense of place is an internally focused process in which disaster-affected people prioritize their activities instead of relying on outside help. It is a relationship-building activity that brings all community members together. By identifying assets, the disaster-affected people identify the human capital that they have in their community.

Hurricane Maria highlighted the importance of memories in reestablishing place, as well as exercising control over meaningful space, the manipulation of that space, and the recreation of some essence of significant past settings in later life. Such acts have important psychological consequences. Specifically, disaster-affected people are motivated to effect these changes in order to discover, confirm, and remember who they are.

15.5 SUMMARY

It is evident that the intensity of disasters like Hurricane Maria will continue to be experienced in the Caribbean. The humanitarian crisis will be characterized by complicated issues of mental health and well-being. In such a situation, the established model of place-based psychosocial support activities can effectively be used to deal with the issues of protecting vulnerable survivors, who often fail to navigate resources and support the remaining marginalized.

Place-based psychosocial support as a strategy has a multi-dimensional impact. It ensures interventions on all three levels (individual, family, and community) and it is accepted by communities faced with stress and traumatic experiences without any stigma attached to mental health issues. Place-based psychosocial support activities are provided by community volunteers in a culturally sensitive manner to strengthen the resources in a community and establish the self-reliance of survivors after a disaster. Psychosocial support activities recognize the strength of a community's inhabitants and it facilitates other interventions. Eventually, integrated approaches may have the potential to rebuild the support systems for vulnerable survivors to facilitate normalization. This approach holistically builds an umbrella of support by bridging gaps between the specific needs of different sectors (for example, livelihoods, housing, health, nutrition, sanitation, and psychosocial support). Furthermore, coherence is actualized between the support system and the affected people.

REFERENCE

Cutter, S. L.; Barnes, L.; Berry, M.; Burton, C.; Evans, E.; Tate, E.; Webb, J. A Place-Based Model for Understanding Community Resilience to Natural Disasters. *Global. Environ. Change* **2008,** *18,* 598–606.

Davis, T. E.; Grills-Tequechel, A.; Ollendick, T. H. The Psychological Impact from Hurricane Katrina: Effects of Displacement and Trauma Exposure on University Students. *Behav. Ther.* **2010,** *41,* 340–349.

Fullilove, M. T. Psychiatric Implications of Displacement: Contributions from The Psychology of Place. *Am. J. Psychiatry.* **1996,** *153,* 1516–1523.

Hobfoll, S. E.; Watson, P.; Bell, C. C.; Bryant, R. A.; Brymer, M. J.; Friedman, M. J.; Ursano, R. J. Five Essential Elements of Immediate and Mid-Term Mass Trauma Intervention: Empirical Evidence. *Psychiatry* **2007,** *70,* 283–315.

McFarlane, A. C.; Williams, R. Mental Health Services Required After Disasters: Learning from The Lasting Effects of Disasters. *Depress. Res. Treat.* **2012,** *2012,* 970194. DOI: 10.1155/2012/970194. Epub 2012 Jul 1.

Norris, F. H. Disasters in Urban Context. *J. Urban Health* **2002,** *79,* 308–314.

Prewitt Diaz, J. O. Integrating Psychosocial Programs in Multisector Responses to International Disasters. *Am. Psychol.* **2008,** *63,* 820–827.

Prewitt Diaz, J. O. Recovery: Re-establishing Place and Community Resilience. *Global J. Community Psychol. Pract.* 2013, *4,* 1–10.

Prewitt Diaz, J. O.; Dayal, A. Sense of Place: A Model for Community-Based Psychosocial Support Program. *Australas. J. Disaster Trauma Studies* **2007,** 2008-1. http://trauma.massey.ac.nz/issues/2008-1/prewitt_diaz.htm.

Saraceno, B. Foreword. In *Fundamentals and Overview of the Handbook of International Disaster Psychology, Vol. 1* Reyes, G., Jacobs, G. A., Eds. Praeger Publishers: Westport, 2006; pp 7–9.

Shalev, A. Y.; Ursano, R. J. Mapping the Multidimensional Picture of Acute Response to Traumatic Stress. in *Early Intervention for Psychological Trauma;* Schneider, U., Ed.; Oxford University Press: Oxford, England, 2003; pp 162–186

Williams, D. R.; Steward, S. L. Sense of Place. *J. Forest.* **1998,** *98*, 18–23.

POSTLUDE:
LESSONS LEARNED: ROUTE MAP FOR THE FUTURE: PSYCHOSOCIAL SUPPORT PROGRAM PLANNING AND DEVELOPMENT

JOSEPH O. PREWITT DIAZ

The objective of this book was to provide an overview of programs that have worked in the past during major disasters, school fires, and other catastrophic events in schools, and the impact of psychosocial support interventions during health crises. The authors are practitioners who have spent years refining methods of community approach. Their interest has been to assist affected people to reestablish their place. As we all talked about the sections of the book and what would be some important takeaway, seven main ideas came to the fore. This final section of the book provides seven nuggets of knowledge that the reader should commit to memory.

EVERY PROGRAM HAS A GESTATION PERIOD

The expectation that a psychosocial support program is going to begin immediately is not appropriate. Every disaster brings different groups of responders in the acute and reconstruction phases. For GOs and INGOs who are prepared to serve for a short period of time (3 months to 1 year), it is best to plan interventions within the existing government structures. For organizations planning to have psychosocial support programs during reconstruction to development, there is a need for careful planning interventions with the primary stakeholders. This gestation period (often as long as 6 months) allows for the careful planning, building the capacity of personnel on the ground, identifying the language of distress, severity, duration and what the community does to feel better and develop materials that are contextualized, are linguistically and culturally appropriate. All aspects of the program from planning to evaluation is conducted by local resources.

PSYCHOSOCIAL INTERVENTIONS SHOULD MATCH THE LEVEL OF MOVEMENT OF AFFECTED PEOPLE TOWARD THE REESTABLISHMENT OF PLACE

People mourn their losses and begin to bond with the new place. Rituals from the old place and rituals from the new place are essential to the process of psychological rebuilding. The paradigm shift from formal mental health interventions in the health post or clinic to formal interventions by outreach teams or nonformal interventions by community volunteers may be the best way to reach the unreachable. Canvassing the neighborhood and engaging the survivors in community mapping may lead to more inclusive of persons that would otherwise not be counted. Community volunteers that manage simple tools like psychological first aid are invaluable on the road to recovery.

COORDINATION AND COMMUNICATION WITH INTERNAL AND EXTERNAL STAKEHOLDERS PRIOR TO A DISASTER FACILITATES REACHING THE DISASTER AFFECTED QUICKLY

Coordination needs to take place with the other partners, non-health related organizations, and emergency management. Whereas, possible protocols should exist in the country's emergency response plan outlining the roles of different partners in psychosocial support activities as part of the preparedness phase of disaster management. Well articulated response protocols and long-term planners embedded in the immediate response teams will guarantee uninterrupted services among the different phases of disaster reconstruction to development.

UTILIZE A COMBINATION OF INTERVENTIONS TO ASSURE THAT A COMMUNITY HAS PARTICIPATED IN IDENTIFYING THEIR EMOTIONAL, PSYCHOLOGICAL, AND SOCIAL NEEDS

Use quantitative and qualitative methods for recording language of distress. Most people will be distressed because they are experiencing loss of place, familiarity, and attachments (individual and collective). The signs of distress are related to disorientation, alienation, and nostalgia. The community should be involved in defining its language of distress and devising the potential interventions it needs to resolve the disaster-related stress. All intervention should be timely and experiential. They should be linguistically, contextually, and culturally appropriate. All intervention should be available to all groups within a community and should reduce any type of stigma.

SIMPLE, ACCURATE, AND TIMELY INFORMATION REDUCE DISASTER-INDUCED DISTRESS

Information dissemination to the primary stakeholders has to be clear, simple, and easy to visualize. All information must be contextualized. To simply translate a handout to the target language does not guarantee that the primary beneficiaries will understand the messages (see Chapter 13 for Best practices). The higher the amount of clear information that addresses basic needs, security, and belonging needs, the faster people take charge of their own recovery.

INTERVENTIONS MUST BE SIMPLE, IMMEDIATE, EXPERIENTIAL, AND SIMPLE

Community-based interventions by health promoters and volunteers are effective. One evidenced informed tool is psychological first aid. If additional individual needs arise then referral to the Health Post is in order.

ASSIST LOCAL GOVERNMENTS TO INSTITUTIONALIZE PSYCHOSOCIAL SUPPORT PROGRAMS

Psychosocial support is a simple program that can be easily institutionalized by the communities and local agencies. The ultimate goal should be to provide a sense of safety, security, calmness, and stability. Calmness and stability allow people to develop an intimate knowledge of their settings and to develop trusting relations with place and with each other, reintegration of the physical and psychological community, and recovery of home, family, friends, and church.

INDEX